# Sign Out
# Science

Silvana Carletti

Suzanne Girard

Kathlene R. Willing

# Sign Out Science

*Simple hands-on experiments using everyday materials*

Pembroke Publishers Limited

© 1993  Pembroke Publishers Limited
538 Hood Road
Markham, Ontario L3R 3K9

**Canadian Cataloguing in Publication Data**

Carletti, Silvana
  Sign out science

Includes bibliographical references and index.

ISBN 1-55138-000-5

1. Science – Experiments. 2. Scientific
recreations. I. Girard, Suzanne, 1950-
II. Willing, Kathlene, 1937-   . III. Title.

Q164.C37 1993      502'.8      C93-093311-7

Editor: Joanne Close
Design: John Zehethofer
Cover Photography: Ajay Photographics
Typesetting: Jay Tee Graphics Ltd.

Printed and bound in Canada
9 8 7 6 5 4 3 2

# Contents

# Acknowledgements

As a result of encouragement from colleagues over the last few years, the original concept of portable science activities in bags has evolved into Sign Out Science. Leon Thompson, Toronto Board of Education Superintendent, supported the concept in its initial stage. Many of the bags found in this book have been tested in numerous schools in the Toronto, North York, and City of York Boards of Education, as well as the Cuthbert Moore Primary School in Barbados.

In writing this book we were pleased to have had the invaluable insights of the following educators: from the Toronto Board of Education, Denis Cooke — Coordinator of Science, Ann Moriarity — Principal, and Cheryl Cox — Teacher; from the Metropolitan Separate School Board, Raymonde Taillefer — Teacher, and Pat Cianfaranni — Teacher. They brought their consultative, administrative, and classroom expertise to bear by reading the manuscript, contributing practical suggestions, and lending encouragement to our project.

Appreciation is extended to Michiomi A. Kabayama, D. Sc. for his scientific advice and culinary contributions.

We were grateful to Eureta Bynoe — Principal, staff, and parents at Earlscourt Public School in Toronto. They implemented a Sign Out Science program in their school and extended it to their computer-networked partner, Cuthbert Moore Primary School, Barbados, during a school-exchange visit. Barbara Howard — Principal, Patricia Saul — Teacher, and other staff responded favorably to Silvana's student workshops on Sign Out Science.

To Dr. Allan MacKinnon, Faculty of Education, Simon Fraser University, we extend a heartfelt thanks for his initial enthusiasm and on-going interest in our endeavors. We have come full circle through his introduction.

# Foreword

The Sign Out Science story began in the summer of 1989 while Silvana attended the Ministry of Education Additional Qualification Course, Primary/Junior Science. The course instructor, Allan MacKinnon, was a member of the Faculty of Education at the University of Toronto. Silvana had always been interested in science. Her classrooms and libraries featured science centers, but she found, to her dismay, that there was never enough time to do all the activities.

Challenged by Allan's assignment to create an innovative project and by her desire to promote science, she was inspired to develop science experiments in bags that could be signed out of her school library. Her original "Science-in-a-bag" project met the challenge.

Suzanne, a colleague in the same course, was writing *Learning Together: Computer Integrated Classrooms* with Kathlene that summer. When Suzanne heard about Silvana's ideas and experience, she asked her to consider co-authoring a library-resource book for teachers once *Learning Together* was published in September, 1990.

In late summer of 1990, Suzanne, Kathlene, and Silvana began to write. One of the topics in their book was Silvana's "Science-in-a-bag" project, which she had implemented in her library. The science bag concept evolved into Sign Out Science© and was included in *The Library/Classroom Connection*, published in September, 1991.

The Sign Out Science section of *The Library/Classroom Connection* attracted educators' attention. In response to their enthusiasm about integrating hands-on science activities across the curriculum, Silvana, Suzanne, and Kathlene extended the Sign Out Science concept into the book you are now reading.

# Introduction

*Sign Out Science*, while it is an excellent source of ideas, is more than another collection of science activities for children. It reflects a recent movement in educational thought that connects children's lives in elementary school with a much wider experiential and cultural base — that of the family. Children's education is intimately related to the fabric of activity and relationships they experience at school with teachers and fellow students, and at home with siblings, parents, and grandparents.

Learning about science takes place in a rich environment in which the expression and exchange of ideas about our natural world is valued. In such an environment we don't simply tell children that their ideas are "right" or "wrong"; we ask about their *reasoning*, acknowledging that they are sensible people and their ideas are worthy of careful consideration. Our task as educators is to discover the reasoning and rationality of children, and, on the basis of that foundation, design further learning experiences. In science, hypotheses and mental representations about why things occur provoke the testing and refining of understandings through "doing." What is required is not merely a set of stimulating events, such as those guaranteed by the activities in this collection, but an audience — a culture — for young people to sound out their thinking. Children's conceptual growth is thus inextricably bound to their own interests and aspirations, the social relationships that help to form these interests and aspirations, and the specific contexts of inquiry at hand. Their knowledge of scientific facts and concepts emerge from imaginative and playful exploration with material, as well as the talk that surrounds and propels further play and inquiry. Teachers and parents alike help to sustain children's interests, self-esteem, and attitudes toward life-long learning when they understand and appreciate how the young child makes sense of things. When home and school work together in concert there is a much greater chance that this care and understanding will occur. That children come to see themselves as people who are *capable of making sense* is foundational, then, to their intellectual, social, and emotional growth.

How can we, as parents and teachers, help children to learn about science? First, we must understand that science is much more than a collection of established facts and theories. It is true that scientific inquiry relies on principles, models, theories, concepts, and procedures that are conventions of the scientific community, but these things are not self-evident in a world of "discovery." They are the products of human invention and ingenuity, and they draw their meaning when they are taken and applied to interesting problem situations. Further, as alluded, science educators are beginning to see that learning about science is much more a function of social activity than individual, and we understand children's learning in science now as being dependent on social relationships and the occasion to express current understandings and representations. Indeed, we believe the very act of articulating an explanation through talk or writing helps learners to sort out their understandings.

We must also understand that learning science involves much more than memorizing facts. Scientific understanding is embedded in a process of thinking about questions, such as, *"Why does this happen?"* or *"What would happen if...?"* When children ask these questions for themselves, all the better. We tend to think that purposeful learning occurs when children have their "hands on" and "minds on" activities, as well as a reason to engage in them in the first place. We help them sustain *curiosity* when we model the same attitudes and quest for understanding ourselves.

I remember teaching an Additional Qualifications course in elementary science education to a group of teachers during the summer of 1989 at the Faculty of Education, University of Toronto. What it meant for me and the participants, including Suzanne and Silvana, was that we were all "becoming" teachers of science to young people. We were *alive* as a group then, and we've gone on to grow and learn in ways that hold us together even stronger than before. Who would have guessed that this book, among others, would come from the seeds that were planted then? Many of us have used *Sign Out Science*, thanks to the initial ideas Silvana generated in her project. I have used it on many occasions — in Ontario and, since moving to Simon Fraser University, in British Columbia. Each time it has been an overwhelming success. *Sign Out Science* seems to be the magical ingredient that helps to develop wonderful collaborations between school and community. I'm convinced by my experiences that *Sign Out Science* is a winner among children, parents, teachers, librarians, and school administrators alike. This book is a wonderful introduction to a strategy for promoting elementary science education that seems to work time and time again and reflects sound educational theory about the way children learn science.

*Allan MacKinnon, Ed. D.*
*September 1992*

# 1

## Sign Out Science

> The application of science is important in a
> changing world because it is a transferable
> skill. The content will change, but that skill
> is what they are left with.
>
> *Dr. David Suzuki*

## Why Sign Out Science?

Sign Out Science bags make science fun and encourage positive attitudes. An active, hands-on extension of school work, they contain all the materials and equipment needed to carry out simple experiments and may be signed out in the same way that books are signed out of libraries.

Simple, open-ended experiments in the bags allow parents, teachers, and students to work together. When used in the home, they provide parents and children with opportunities to share quality time together while also allowing them to formulate and test hypotheses, and discuss conclusions. By testing scientific facts, students develop skills including those of awareness, knowledge, and critical thinking. Open-ended questions at the end of the experiments foster additional experimentation and further research, and provide the opportunity to apply the knowledge gained. Sign Out Science bags hold a fascination for students of all ages and provide theme-related, hands-on investigative activities that are enjoyable for both children and their parents.

Conducting the activities in the bags provides students with a means of manipulating and questioning aspects of their environment. Intent on bringing meaning to their activities, students are keen to develop competence. They learn best in authentic situations, such as those found in Sign Out Science, where hands-on investigations lead to meaning. When questioning is transferred from adults to students, they become actively involved and take command of their learning. Sign Out Science bags stimulate thinking, heighten students' interests, and provide fun.

In addition, the experiments in the book are pedagogically sound. As an example, the information in the *Life swingers* activity introduces the students to a swinging pendulum. Through this introduction, they gain the knowledge of how a pendulum works, build concepts around time and motion, and make generalizations about how changes in the pendulum affect its motion. The hands-on approach engages them in an activity that helps them use this new information in a meaningful way and allows them to bring their own perceptions to the activity. It is important that students have the time and the space to think about their current observations in the context of their past experiences.

Other *knowledge concepts* developed in the bags are:

| | | |
|---|---|---|
| change | community | conservation |
| energy | growth | life |
| matter | space | technology |

A bag such as *Mini-composter* allows students to observe the decomposition of objects and develop appropriate, positive attitudes about concern and care for the environment. Consequently, students are more likely to use the knowledge and skills they gathered in the activity in responsible ways. A natural extension of this activity involves the research and implementation of a recycling program in their school.

Other attitudes developed in the bags are:

| | | |
|---|---|---|
| concern for accuracy | consideration | cooperation |
| curiosity | for others | perseverance |
| respect for living things | open-mindedness | safety- |
| thoroughness | risk-taking | mindedness |

The interconnected process skills of scientific inquiry are also important. These processes, ranging from simple to complex, can be practised at all grade levels. For example, *Slippery surfaces* involves students in looking at friction as a force that resists motion. This requires students to develop and use skills in well-devised, logical, and intuitive ways. As they control variables, use equipment, and seriate their observations, students begin to understand the properties of the materials they are using.

Scientific inquiry is developed through:

| *simple process skills* | *complex process skills* |
|---|---|
| classifying | experimenting |
| communicating | controlling variables |
| inferring | hypothesizing |
| measuring | interpreting |
| observing | making models |
| predicting | using equipment |
| seriating | |

Knowledge, attitudes, and skills are developed in the bags, but some aspects are more predominant than others. Knowledge is linked to the materials, the process, and the application of the general concepts in everyday life. The intrinsic nature of the bags incorporates positive attitudes towards risk-taking and curiosity. Observing, communicating, and inferring are the skills most inherent in the bag activities.

The bags in Chapters Two and Four focus on various learning opportunities. As the students explore the bags, they move through the three stages of skill development — awareness, practice, and transfer — that are built into the bag's activities. The basic activity will heighten awareness and provide practice. The extension activities provide more practice and allow for transfer of concepts to other tasks. Each time they explore the bags, students' level of skills are at different stages of development depending upon their age, maturity, and science experience. Thus they can choose to use the same bag several times throughout the year. This should be kept in mind when developing bags and observing their use.

## Collaborative planning

Although it is possible to plan and develop a Sign Out Science program independently, a collaborative approach provides a richer program. Within the context of resource-based learning, a wide range of human, material, and physical resources exist within the school and community. Colleagues, high school teachers, consultants, teacher/librarians, students, and parents can share in the planning. Kits, films, videos, software, and books may be used. Places to visit include computer labs, resource centers, field-trip sites, and public libraries. For general collaborative planning, *The Library/Classroom Connection* details the process, including brainstorming, webbing, and completing a planning guide.

In addition to the general benefits of collaborative planning for Sign Out Science, there are also specific advantages. In an area that is changing rapidly, having someone with science expertise involved is comforting. People from different backgrounds generate a greater wealth of ideas, and when it comes to putting the bags together, the workload is shared and more bags are developed.

It is important for everyone at each stage of the planning process to ensure that all components of science — Earth and Space, Life, Physical, and Technology — are covered.

Earth and Space topics include:
- time
- seasons
- the universe
- relationships
- cycles
- weather
- properties and functions of particles
- appropriate technology

Life topics include:
- adaptations
- living things
- the environment
- life cycles
- classification of living things
- appropriate technology

Physical topics include:
- energy
- structures
- changes
- states of matter
- properties of materials
- appropriate technology

The following overview details each of the twenty-four Sign Out Science bags — suggested materials, area of science focused on, and the four major skills developed in the bag's activities.

SIGN OUT SCIENCE AT A GLANCE

| Title | Page | Suggested Materials | Area | Skills |
|---|---|---|---|---|
| Life swingers | 35 | • roll of Life Savers®<br>• thread (at least 1 m)<br>• tape<br>• measuring tape | Earth and Space | • observing<br>• measuring<br>• constructing<br>• experimenting |
| Balloon whizzer | 38 | • balloon<br>• straw<br>• string (3 m)<br>• masking tape | Earth and Space | • observing<br>• measuring<br>• constructing<br>• communicating |
| Color it hot | 40 | • 2 plastic glasses<br>• sheet of white paper<br>• sheet of dark paper<br>• pencil<br>• thermometer<br>• clear tape<br>• graphic organizer | Earth and Space | • observing<br>• measuring<br>• recording<br>• experimenting |
| Shifting time | 43 | • 2 small plastic bottles<br>• sand<br>• pen<br>• masking tape<br>• small cardboard disc | Earth and Space/<br>Technology | • predicting<br>• inferring<br>• measuring<br>• making a model |
| Dry bones | 46 | • clean, dry chicken and fish<br>  bones<br>• magnifying glass and<br>  different magnifiers | Life | • observing<br>• classifying<br>• predicting<br>• experimenting |

SIGN OUT SCIENCE AT A GLANCE

| Title | Page | Suggested Materials | Area | Skills |
|---|---|---|---|---|
| Take a breath | 49 | • plastic, 2 L bottle<br>• Plasticine<br>• straw<br>• balloons (1 small, 1 large)<br>• 2 elastic bands (1 small, 1 large) | Life | • observing<br>• constructing<br>• predicting<br>• inferring |
| Germinating seeds | 51 | • small plastic bag<br>• paper towel<br>• seed, such as bean, pumpkin, or corn<br>• calendar<br>• pencil | Life | • observing<br>• constructing<br>• predicting<br>• measuring |
| Mini-composter | 54 | • clear plastic bag<br>• soil<br>• small piece of newspaper<br>• metal cap or tab<br>• small piece of styrofoam<br>• pencil<br>• time line | Life/Technology | • observing<br>• constructing<br>• experimenting<br>• predicting |
| Colorful black | 57 | • coffee filter<br>• clear plastic glasses<br>• water-soluble marker | Physical | • experimenting<br>• inferring<br>• hypothesizing<br>• predicting |

SIGN OUT SCIENCE AT A GLANCE

| Title | Page | Suggested Materials | Area | Skills |
|---|---|---|---|---|
| Slippery surfaces | 60 | • spring scale<br>• string<br>• masking tape<br>• measuring tape<br>• graphic organizer | Physical | • observing<br>• measuring<br>• inferring<br>• controlling variables |
| Detecting prints | 63 | • magnifying glass<br>• talcum powder<br>• black construction paper | Physical | • observing<br>• classifying<br>• experimenting<br>• communicating |
| Speedy marbles | 65 | • marbles<br>• 2 small blocks of wood<br>  or 2 small books<br>• masking tape | Physical/Technology | • observing<br>• constructing<br>• measuring<br>• experimenting |
| Bobbing apples | 83 | • plastic knife<br>• plastic container<br>• pencil<br>• paper | Earth and Space | • experimenting<br>• communicating<br>• interpreting<br>• measuring |
| Seedy fruits | 86 | • magnifiers<br>• plastic knife<br>• pencil<br>• paper | Life | • classifying<br>• communicating<br>• observing<br>• interpreting |
| Seed rattles | 88 | • 3 small containers<br>• apple, orange, and pumpkin<br>  seeds | Physical | • communicating<br>• controlling variables<br>• hypothesizing |

# SIGN OUT SCIENCE AT A GLANCE

| Title | Page | Suggested Materials | Area | Skills |
|---|---|---|---|---|
| Seed rattles (cont'd) | | • graphic organizer<br>• pencil<br>• magnifier | | • observing |
| Apple juicer | 91 | • small hand grater<br>• plastic knife<br>• wooden craft stick<br>• small plastic container<br>• 30 cm by 30 cm piece of cheesecloth<br>• large piece of waxed paper | Physical/Technology | • communicating<br>• making a model<br>• manipulating equipment<br>• observing |
| Acid snow | 94 | • 35 mm film containers<br>• labels<br>• pH paper<br>• swimming pool pH tester | Earth and Space | • communicating<br>• observing<br>• measuring<br>• experimenting |
| Frozen fruit and veggies | 97 | • paring knife<br>• magnifying glass<br>• plastic bags | Life | • inferring<br>• observing<br>• experimenting<br>• hypothesizing |
| Icy icicles | 99 | • clear plastic bottles<br>• large needle<br>• food coloring<br>• string<br>• tape | Physical | • communicating<br>• experimenting<br>• hypothesizing<br>• making a model |

# SIGN OUT SCIENCE AT A GLANCE

| Title | Page | Suggested Materials | Area | Skills |
|-------|------|---------------------|------|--------|
| Snow gauge | 101 | • clear plastic bottle<br>• masking tape<br>• permanent marker<br>• scissors | Earth and Space/Technology | • experimenting<br>• measuring<br>• manipulating equipment<br>• interpreting |
| Around and around it goes! | 104 | • 2 small light-colored balloons<br>• coin<br>• marble | Earth and Space | • observing<br>• inferring<br>• communicating<br>• experimenting |
| Pinch and lift | 106 | • levers (e.g., chopsticks, tweezers, ice tongs)<br>• balloon<br>• marble<br>• penny | Life | • observing<br>• interpreting<br>• seriating<br>• inferring |
| Easy moves it | 109 | • piece of cardboard<br>• marbles<br>• lid from a box (at least 20 cm by 20 cm) | Physical | • experimenting<br>• inferring<br>• making a model<br>• observing |
| Go car go! | 112 | • 1.5 volt motor<br>• 1.5 volt battery<br>• piece of cardboard<br>• 6 elastic bands<br>• scissors<br>• white glue<br>• 2 soup cans (preferably broth)<br>• 4 washers | Physical/Technology | • experimenting<br>• making a model<br>• predicting<br>• communicating |

## Science Webs

A science web provides a quick visual check of the overall distribution of the bags. The bags developed for the *Ice and snow* theme in Chapter Four are represented on the following web.

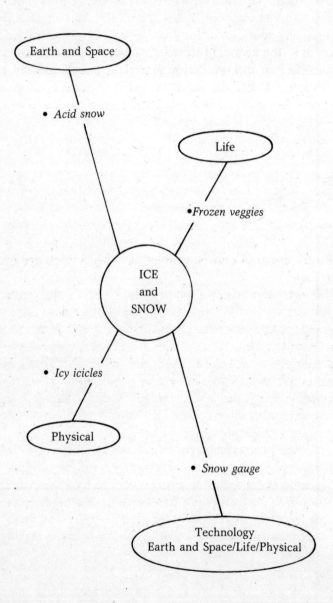

# Development

Although many bags have been collaboratively planned and developed for theme integration, bag ideas can come from anywhere. Teachers often take advantage of those precious teaching moments when students display curiosity and interest. Objects from home, literature, films or videos, trips, or neighborhood walks may spark ideas for another bag. Background information related to the kernel of the idea can be found in science resource books, fiction and nonfiction materials, encyclopedias, computer software, CD-ROMs, laser- or video-discs, and dictionaries.

Once the research is done, the development of a prototype of the bag begins. It is important to select an activity that is suitable to students and related to a theme or unit being studied. The activity can then be narrowed down to one concept; concentrating on more than one idea may cause confusion for students. Write a rough draft of the necessary materials and procedure in clear, easy-to-understand language and consistent format. A useful way to help students organize information, such as their observations, is to provide them with graphic organizers. These can take the form of charts, graphs, or webs which are included in the bags.

To test the prototype, gather the materials and do the activity following the draft directions. Anything that does not work, such as steps in the procedure, confusing phrases or words, inappropriate materials or observations, and unexpected outcomes can be changed at this time. Add extensions that further the bag's single concept and take the student one step beyond the bag. Look for connecting links to other subject areas or themes. A good way to expand a collection of Sign Out Science bags is to share research notes and prototypes with other educators.

Once the prototype has been tried and changes have been made, a final revised copy of the instructions can be written. As well, a separate sheet detailing a series of Sign Out Science tips, along with a brief resource list, can be mounted on the reverse side of the instruction card in the bag. The following is an example of a Sign Out Science Tips sheet to go home in the bag for parents, guardians, and students.

# Sign Out Science Tips

**Remember to ...**

- read the whole activity before starting
- predict what will happen
- follow directions carefully
- repeat the activity and compare the results
- discuss and draw or write about what happened
- think of changes to the activity and try them out.

**Look for these ...**

*An Early Start to Technology* by Roy Richards
*Science Express: An Ontario Science Centre Book of Experiments*
*Power Magic* by Alison Alexander
*175 Science Experiments to Amuse and Amaze Your Friends* by Barbara Walpole

The assembly of the Sign Out Science bag involves:
- placing the materials and instruction/tips card in a labelled plastic bag
- including a small notebook for parents and children to share discoveries, questions, and suggestions
- adding a literature component by putting related picture books, poetry, or other fiction in the bags
- attaching a library pocket and card for easy signing out.

Once the prototype has been tested and the final bag has been assembled, accumulate materials for a number of bags. Popular bags require sufficient quantities on hand for quick turnovers. If materials are not available to refurbish them, students will stop borrowing the bags. Therefore, recycled and inexpensive materials should be on hand at all times. Gathering materials can be made easy when everyone — students, parents, staff, and local businesses — becomes involved.

Tips for collecting materials:
- request donations of small items, such as marbles, buttons, corks, nails, and vials
- buy inexpensive items, such as toothpicks, balloons, balls, straws, and string
- save empty plastic bottles, jars, and containers, including those from yogurt, margarine, and 35 mm film.

Some materials may be difficult to find. Items that closely approximate the qualities of the object can be substituted. For example, similar-sized cans may be used for the ones in Bag Four on Technology, *Go car go*, found in Chapter Four. The rule to follow is this: *substitution is okay as long as the results are not affected.* However, if after testing the substitute, the overall results are affected, the activity must be redeveloped or rewritten.

## Implementation

Once the bags have been assembled, Sign Out Science is ready to be implemented. Perhaps the best way to introduce the new program to students is to hand out the bags and let them "jump in." While students are busy with the activities, the teacher can circulate and encourage observation and questioning. When the investigations are complete, the bags can be put together and collected.

At this time, a class discussion about routines would be helpful. Students can focus on the need for ways to handle borrowing and refurbishing the bags. Steps for signing out and returning bags can be established.

Consumable materials will need to be replaced. These can be sorted and stored in bins or boxes. The bins can be a part of a Sign Out Science center in the classroom. In the same area, the bags and a sign-out board can be displayed. Related resources can also be placed where they will be accessible to students and parents. Student monitors and parent volunteers can help refurbish the bags.

Parents play a role in Sign Out Science so it is important to make contact with them as part of the program's implementation. Students can take home their introductory bags to show their parents. An accompanying letter could be one of two types. It could explain the routines, describe the parent's role, and request assistance, or it could be an invitation to attend a Sign Out Science Evening. The following is a sample note to parents that can easily be adapted.

---

Dear Parent,
Your child is embarking on a new science adventure — Sign Out Science. What makes this even more exciting is the fact that you will have a chance to share science discoveries with your child.

This bag is just one of a variety of ready-made science bags we have on hand for your child to bring home. The easy-to-follow instructions and handy materials are included in each bag. All you have to do is have fun learning with your child. You will soon receive an invitation to attend our Sign Out Science Evening, which will be an opportunity to find out more about the program.
Sincerely,
S. Carletti

---

## Parental involvement

Generally, children whose parents or care givers are involved in their education are more likely to succeed in school. Knowing this, schools regularly include a parent component in many programs. Parental involvement enhances the success of a Sign Out Science program, whether it is on a classroom, division, or school-wide basis. Some parents can provide expertise about science topics in the bags. At school, parents are valuable volunteers who help with borrowing routines and refurbishing bags. Most important, parents at home assume the key role of facilitator by investigating bags with their children.

Parents become involved in Sign Out Science when they know it is a nonthreatening way to help their children learn. The purpose of the program is to promote positive attitudes to science through conducting enjoyable science activities. When parents understand this, they view it as spending meaningful time with their children. Many parents also derive satisfaction from contributing as resource people and volunteers.

Parents appreciate on-going contact with their children's school and enjoy school functions where refreshments are served and they can get to know one another. As soon as the Sign Out Science program has been introduced to students, word about it can be sent home. An effective way to initially involve parents is to have students take home a bag they have used in class, along with a personalized invitation to attend a Sign Out Science evening at school. In addition to presenting a variety of bags, the evening would include a discussion of the program's purpose, benefits, and implementation.

## An introductory evening

Organizing a social event is an effective way to attract parents to the school. They can be invited, along with their children, to attend a special event where the Sign Out Science program is introduced. As with any get together, participants should be welcomed as they arrive. Name tags and introductions ease conversations. As well, refreshments contribute to the hospitality. If it is possible, a simple, casual supper makes it convenient for families to attend early in the evening. Juice, tea, coffee, and cookies are always appreciated. A visually stimulating display could be available for browsing throughout the evening. Books, photos, posters, students' work, and videos arouse curiosity and spark interest.

Once the guests have arrived, the host teacher outlines the agenda. Other staff members, including the principal, may have a role to play. A guest speaker could briefly address a broad science issue to "set the stage" for the evening. The host teacher would describe the Sign Out Science program in general, and its classroom implementation in particular. It is important that parents know the borrowing and refurbishing routines, and crucial that they recognize their role as facilitator and how to work with their children at home.

The bulk of the evening should be spent having families explore a variety of Sign Out Science bags. This provides a good opportunity for teachers to circulate, model the facilitator's role, and encourage use of the resource display.

After the hands-on activity, refocus the participants' attention for a brief period before they go home. Now that the parents understand the program's goals and content, elicit their on-going support. When asking for volunteers, state job expectations clearly and specifically, along with the time commitments required. In addition to volunteering to help refurbish and to circulate bags, parents can give demonstrations and explain their science-related jobs. Women and minorities with careers in science and technology should be encouraged to participate as role models. Have a volunteer list handy for signing before parents leave. Whether planned monthly or once a term, Sign Out Science evenings develop positive attitudes to science among those in attendance.

The overall approach to Sign Out Science may differ from many parents' own early learning experiences. Therefore, encouragement and support helps them maintain involvement in the program. They may be interested in a classroom newspaper with a Sign Out Science column. School newsletters can acknowledge parents' contributions. Sign Out Science Evenings can be held regularly. Parent resources and volunteers can be invited to an end-of-year celebration. On-going suggestions for parents on how they can help their children and become catalysts to learning are useful. The Sign Out Science Tips sheet on page 24 is one reminder, while a Sign Out Science flyer that can be sent home and posted in a prominent place is another. The flyer might include pointers such as:

- encourage your child to read the entire activity before starting
- promote the discovery process by asking questions
- answer questions with questions
- suggest experiments be repeated several times and compare results
- support creativity in and ownership of investigation
- urge your child to find applications in his or her daily life.

## Evaluation

As noted before, Sign Out Science is an extension that is meant to be an enjoyable learning experience. While there are other areas of the classroom science program that are formally evaluated, Sign Out Science, by its nature, lends itself to informal observation of students. Evaluation can be based on student response and student/teacher observation. Although the bags are signed out and taken home, they should not be considered as traditional homework to be marked and corrected. The focus is on observing changes in attitudes toward science, and development of thinking skills. Informal assessment tools, such as response journals or logs, can be used to look for deeper meaning in science, along with observational checklists to document and understand students' involvement in the program.

The purpose of using Sign Out Science journals/logs is to provide students with a way to keep records of and responses to the bags they have used. Children and parents having English as a second language may be assisted in this writing task by teachers, specialists, parents, and students. Science journals/logs are a means of reflecting on experiences and can be used as references for teacher observations. The journals/logs can take a variety of forms. Computer software, such as a word processing package, separate notebooks, or science notebooks can be used. Photocopied sheets, designed specially for the Sign Out Science program, may be kept in folders. The format for a record sheet might be: student's name, title of bag, date used, and a section for comments.

Teachers in primary grades may wish to design a questionnaire to stimulate responses. Sample questions include:

What science bag did you use?
Why did you choose that one?
When did you use the science bag?
What did you think would happen in this bag?
What did happen?
How did you like the bag?

Students use language arts skills to express scientific experiences when they maintain response journals. Some of their responses — how they felt about using the bag and their likes/dislikes — come from the affective domain. Writing entries becomes a meaningful process that encourages students to work within an integrated learning environment.

A new dimension to response journals is to code journal entries. Students' responses reflect their knowledge, skills, and attitudes. In the classroom, the teacher and students can create science-style symbols or codes to match identifiable parts of their thinking process. These symbols are used by the teacher and students to code students' entries. For example, a magnifier may be a code for questioning that seeks deeper meaning. Each time a student write entries, such as, "I wonder what would happen if...?" or "What else might make this...?" a small magnifier is drawn in the margin next to the matching sentence. By glancing at accumulated codes, an overview of a student's thinking strategies becomes evident.

Coded thinking skills and strategies to look for in Sign Out Science journals/logs are:

 making predictions

 comparing

 contrasting

 recognizing cause and effect

 evaluating and forming judgments

 activating prior knowledge

 relating to personal experience

 questioning for deeper meaning

Adapted from Knight, J. (1990). "Coding Journal Entries." *Journal of Reading*, 34, 42–47. International Reading Association.

An observational checklist is an informal evaluation instrument that can be used by both students and teachers. It is based on information gathered from classroom observation, interviews, and science journals/logs. A self-evaluation component is valuable during student conferences. Students and teachers bring completed evaluation forms to the conference, compare notes, and discuss their findings. A completed observational checklist is also an easy way to survey changes in science knowledge, attitudes, and skills, and is helpful as a discussion point during parent-teacher interviews. The following serves as a sample checklist.

# SIGN OUT SCIENCE CHECKLIST

Student name _____

Observation period _____

| Student Involvement | Self | Teacher |
|---|---|---|
| • signs out bags on a regular basis | | |
| • follows instructions | | |
| • predicts outcome before activity | | |
| • chooses bags from a variety of science areas | | |
| • maintains a science log or journal | | |
| • recognizes connections among bags | | |
| • identifies applications in everyday life | | |
| • uses problem-solving techniques | | |
| • develops extensions | | |
| • uses scientific vocabulary | | |
| • discusses experiences with bags in class | | |
| • maintains a science journal/log | | |

Comments: _____

_____

# SCIENCE WEB

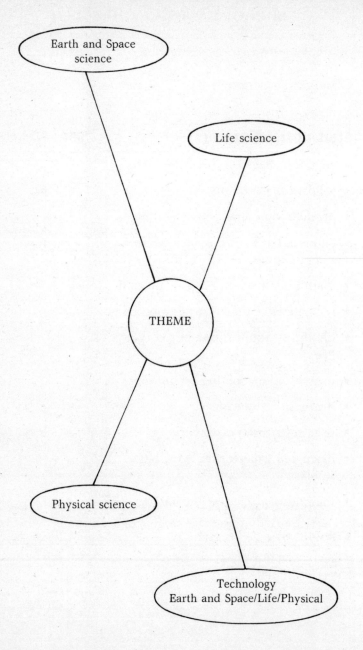

Earth and Space science

Life science

THEME

Physical science

Technology
Earth and Space/Life/Physical

# 2

# General activities in bags

The previous chapter explained the Sign Out Science program. In this chapter, a number of Sign Out Science bags are detailed for class use and to serve as models.

The *Overview* indicates that the bag explores a general topic. The bag's title, science area, and main skills are listed. The *For your information* section introduces general concepts, explains the activity, and offers practical tips. Under *Materials needed for the bag* is a list of suggested items to include. Feel free to substitute items if they are not readily available and other appropriate items are on hand. Selected vocabulary is presented in *Words to know*; activity extensions for teachers are suggested under *Now try this*; and related fiction and nonfiction material appear in *Look for these*.

The instruction sheet intended for students and parents begins with a list of materials found under *You will need*— most materials need to be included in the bags while others, such as water, scissors, or glass jars, are readily available at home. *What to do* provides step-by-step instructions, and is followed by guided observations in *What to notice*, and extensions in *More to do*.

Although technology is inherent in each of the three science areas — Earth and Space science, Life science, and Physical science — it has been accorded extra recognition, in order to heighten awareness and understanding of its presence and value. Technology involves drawing upon past experiences and applying that knowledge to a design process where the end product may be the creation of prototypes and systems. In order to create, change, and improve these prototypes/systems, students examine the materials, the design, and the use of scientific inquiry, convergent, divergent, and critical thinking. Each set of bags is categorized by its science area. The first three bags in each set are labelled with the science area explored, while the fourth bag also contains a technology focus.

Four main science areas pertaining to the bags are drawn from the recognized science inquiry skills appropriate for elementary students. They include: classifying, communicating, controlling variables, experimenting, hypothesizing, inferring, interpreting, making models, measuring, observing, predicting, and seriating. Although these skills are listed separately, they do not exist in isolation. Instead, they are dependent upon the science content at hand, as well as the purpose and context of students' inquiry. Scientific inquiry is holistic and incorporates critical thinking in an integrated approach that enables students to derive meaning and understanding from their explorations.

# Overview — Life swingers

GENERAL

SCIENCE AREA: *Earth and space*

SCIENCE SKILLS: *Observing, constructing, experimenting, measuring*

## For your information...

Pendulums are bobs attached at the end of a support, such as a ball of Plasticine tied to the end of a string. Pendulums are mounted so that the bobs can swing freely. When a bob is pulled to one side and released, its movement is predictable. A full swing back and forth is one vibration during a given time, and the time it takes is a period. This movement is repeated with slightly less distance being covered each time due to air friction. Finally, the pendulum rests in its vertical position.

The frequency of the pendulum is the number of vibrations in a given time. The length of the pendulum determines the number of times it swings — the longer the string, the slower it vibrates. Neither the size nor the material of the bob has any effect on its movement. In the 16th century, Galileo Galilei noticed the regular motion of swinging objects and discovered that the length of the pendulum's swing did not affect the period or time of one vibration. He proved this by counting the number of his heartbeats it took for the objects to swing back and forth.

This activity shows how to make a simple pendulum by hanging a Life Saver® as a bob at the end of a length of thread. It provides students with the opportunities to rediscover Galileo's findings. Students should be encouraged to count the number of periods in one minute in order to find the number of periods per second. They will discover that it is the length of the string that affects the distance and speed of the swing. Caution students not to eat the Life Savers® until the activity is completed!

Pendulums are all around us in the form of grandfather, grandmother, and cuckoo clocks, playground swings, demolition cranes, trapezes, and metronomes.

## Materials needed for the bag...

- roll of Life Savers®
- measuring tape
- thread (at least 1 m)
- tape

**Words to know...**

pendulum             Galileo             period
swing                bob                 vibration

**Now try this...**

- Regulate the pendulum so it has a period of one second, five seconds, 10 seconds.
- Substitute different objects for the bob and different materials for the thread.
- Research the use of pendulums, such as cranes and swings, in work and play.
- Research Galileo's experiments with pendulums and time.

**Look for these...**

*Five Secrets in a Box* by Catherine Brighton. Fitzhenry & Whiteside, 1987.

*Clocks: Building and Experimenting with Model Timepieces* by Bernie Zubrowski. Morrow Junior Books, 1988.

*Up the Science Ladder: Activity Based Ideas for Teaching Science to Primary Grades* by Lynn Molyneux. Trellis Books Inc., 1988.

*Methods of Motion: An Introduction to Mechanics, Book 1* by Jack Gartrell, Jr. National Science Teachers' Association, 1989.

---

*Science Response Journal*
*Life Swingers*

*What I found out about the experiment is that the more lifesavers you put on the thread the more time's it swings. I thought the most difficult thing was counting how many time's it swung. I think I would use 20 seconds instead of 30 seconds and instead of using lifesavers use some buttons. The experiment, reminded me of a pendulum on a clock.*

*David*

# Life swingers

**You will need:**

- small roll of Life Savers®
- small spool of thread
- watch with a second hand
- measuring tape
- table
- clear tape

**What to do:**

Tie one end of the thread to the Life Saver®.
Tape the other end of the thread to the edge of a table.
Measure the length of the thread from the edge of the table to the top of the Life Saver®.
Pull the candy back about 10 cm and let go.
Count how many times the pendulum swings in 30 seconds.

**What to notice:**

- how the pendulum moves
- the distance the pendulum travels
- the time it takes for the pendulum to stop moving

**More to do:**

Change the number of Life Savers®.
Widen or narrow the swing of the pendulum.
Vary the length of the string.

# Overview — Balloon whizzer

SCIENCE AREA: *Earth and space*

SCIENCE SKILLS: *Observing, constructing, communicating, measuring*

## For your information...

Sir Isaac Newton conducted many experiments about motion and how objects react. His statement that for every action there is an equal and opposite reaction is evident during jet propulsion.

This activity uses a blown-up balloon to demonstrate jet propulsion. When the balloon is blown up and the opening sealed, the air pressure is greater inside the balloon. The pressure is changed when the mouth of the balloon is opened. As a reaction, the released air pushes the balloon in the opposite direction. The balloon's movement is directed along a straight path by connecting it to the string. The distance the balloon travels depends on how it is attached to the straw. For best results, the mouth of the balloon should be parallel to the straw and the string as taut as possible. The next time you are on a jumbo jet or ride a monorail notice how the movement is accomplished.

## Materials needed for the bag...

- balloon
- straw
- string (3 m)
- masking tape
- measuring tape

## Words to know...

air pressure

action

taut

reaction

propulsion

Newton

## Now try this...

- Construct racing cars that are powered by blown-up balloons and follow string routes on the floor.
- Make air-powered boats and race them in a water table or a children's wading pool
- Attach different objects to the balloon to determine the maximum load it can move.
- Research and replicate Boyle's Law and Hooke's Law.

**Look for these...**

*Balloon Science* by Etta Kaner. Kids Can Press, 1989.
*A Balloon Goes Up* by Nigel Gray. Orchard Books, 1988.
*Balloon Building and Experimenting with Inflatable Toys* by Bernie
  Zubrowski. Morrow Junior Books, 1990.
*Science Demonstrations for the Elementary Classroom* by Dorothea
  Allen. Parker Publishers, 1988.
*The Balloon Tree* by Phoebe Gilman. Scholastic, 1984.

# Balloon whizzer

**You will need:**

- balloon
- straw
- string (3 m)
- masking tape

**What to do:**

Attach one end of the string to a door knob.
Thread the other end of the string through the straw.
Attach the string to a chair or table, making sure it is taut.
Inflate the balloon and hold the end of it.
Tape the balloon to the straw in a couple of places.
Let go of the balloon.

**What to notice:**

- how far the balloon travels
- in which direction the balloon moves
- how fast the balloon goes
- how the balloon changes
- what makes the balloon move

**More to do:**

Try a different-sized balloon or different lengths of string.
Place a Life Saver® in the open end of the balloon.
Add another balloon and string and have a race.

# Overview — Color it hot

GENERAL

SCIENCE AREA: *Earth and space*

SCIENCE SKILLS: *Observing, recording, measuring, experimenting*

**For your information...**

The sun is a source of heat and light energy. The sun's radiant energy is absorbed when it comes into contact with solids, liquids, or gases. The amount of heat absorbed depends on color, surface area, texture, and translucency of the object. Dull, opaque, or dark-colored surfaces absorb more sunlight; shiny or light-colored surfaces reflect most of the sunlight. Clear objects, such as windows, let the sunlight enter and heat the air. Dark liquids and solids tend to absorb more of the sun's radiant energy than light liquids and solids.

This activity demonstrates that the amount of heat absorbed in a container depends on its color. Identical containers with the same amount of liquid are used. Wrap dark paper with a flat surface tightly around one glass. Wrap light-colored paper around the second glass. For the extensions, students are asked to change other variables, such as the liquid and the color of the paper, to make comparisons. Filling the glasses from the same container of water ensures that the initial fluid temperature is the same in both glasses. This also demonstrates that solar energy is a method of heating that uses the sun's radiant energy, and explains why solar panels are dark in color.

A graphic organizer to help students collect information may look like the following chart:

TEMPERATURE MEASUREMENTS FOR *COLOR IT HOT*

|  | Time #1 | Time #2 | Time #3 |
|---|---|---|---|
| Temperatures for Container #1 |  |  |  |
| Temperatures for Container #2 |  |  |  |

## Materials needed for the bag...

- 2 plastic glasses
- sheet of white paper
- sheet of dark-colored paper
- pencil
- thermometer
- clear tape
- graphic organizer

## Words to know...

| | | | |
|---|---|---|---|
| energy | temperature | reflect | radiant energy |
| solar panels | translucent | opaque | solar energy |
| liquid | variable | absorb | |

## Now try this...

- Using the same color of paper on both glasses, place one glass on each side of a window.
- Construct a solar heater that will heat an aquarium.
- Compare solar energy to other forms of energy, such as wind, hydroelectric, or nuclear energy.

## Look for these...

*Science Demonstrations for the Elementary Classroom* by Dorothea Allen. Parker Publishers, 1988.

*Urban Ecology* by Jennifer Cochrane. Wayland, 1987.

*Ecology* by Richard Spurgeon. Usborne Science Experiments, 1989.

*Sound and Light* by Terry Jennings. A Templar Book, 1992.

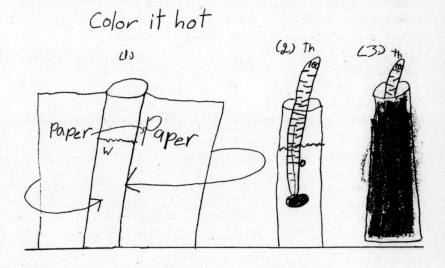

Color it hot

# Color it hot

**You will need:**

- 2 straight-sided glasses
- sheet of white paper
- dark-colored paper
- pencil
- thermometer
- clear tape
- graphic organizer
- container of cold water

**What to do:**

Fill the glasses with equal amounts of water.
Cover the glasses to avoid evaporation.
Measure and record the water temperature in each glass.
Tape the dark sheet of paper tightly around one glass, and the white paper around the second glass.
Set the glasses in a sunny spot.
Measure the water temperature each half hour.
Record at least three sets of measurements on the graphic organizer.

**What to notice:**

- the initial temperature of the water in each glass
- how the water temperature changes each time the measurements are taken
- which glass of water is cooler/warmer

**More to do:**

Add dark food coloring to the water and note the effect.
Vary the amount of water used in both of the glasses.
Use shallower or deeper containers.
Use other liquids, such as vinegar or milk, instead of water.

# Overview — Shifting time

SCIENCE AREA: *Earth and space/technology*

SCIENCE SKILLS: *Predicting, inferring, measuring, making a model*

## For your information...

Egg timers are similar to sand clocks used in the 14th century. They work on the principle of gravity which, on earth, is the force that pulls all objects toward the center of the earth. Sand is made up of small rock particles. It is less affected by humidity than many other materials, hence it moves easily from one container to another under the force of gravity. The time it takes for the sand to drain from a bottle depends on the size of the bottle's hole and the type, condition, and amount of sand.

In this activity, time is measured by how long a quantity of sand takes to flow from one container to the other. This clock prototype has the same-sized hole and the same amount of material employed throughout the activity. It is the extension activities that provide opportunities to change the variables of quantity, hole size, and type of material. Several practices are suggested to allow the students to form an idea of how rapidly the sand flows. When the timing begins in earnest, students need to be able to simultaneously coordinate inverting the clock and checking the watch.

Students can reevaluate, redesign, and improve the prototype by changing the variables. When selecting bottles for the activity, shape and neck size are important. Sand flows better in tapered, or funnel-shaped, bottles. Wider-necked bottles with holes in the caps can be used, but the flat surface does not allow for all the sand to flow out. Smaller-sized holes may impede the sand's flow. Finely ground sand works best, but rice, salt, or sugar can be used as well. The last two materials work best when there is a low level of humidity in the air.

## Materials needed for the bag...

- 2 small plastic bottles
- sand
- small cardboard disc
- masking tape
- pen

## Words to know...

gravity              sand clock              timing              flow

## Now try this...

- Research and design a German sand clock, a composite sand timer that gives readings on each quarter-hour.
- Develop and implement an original time system based on the sand clock.
- Use the sand clock as a timepiece to determine how long certain tasks take.

## Look for these...

*Clocks: Building and Experimenting with Model Timepieces* by Bernie Zubrowski. Morrow Junior Books, 1988.

*Looking Inside: Machines and Constructions* by Paul Fleisher and Patricia Keeler. Atheneum, 1991.

*An Early Start to Science* by Roy Richards. A Macdonald Book, 1987.

*An Early Start to Technology* by Roy Richards. A Macdonald Book, 1991.

*Dropping in on Gravity* by Ed Catherall. Chrysalis Publications, 1988.

*The Scientific Kid: Projects, Experiments, Adventures* by Mary Carson. Harper Collins, 1989.

# Shifting time

## You will need:

- 2 small plastic bottles
- dry sand
- small cardboard disc
- masking tape
- pen
- watch

## What to do:

Fill one of the bottles with sand to about 3/4 full.
Turn the other bottle upside down and place it on the bottle with the sand.
Make sure the openings of the bottles are matched and touching.
Tape the two bottles together where they join to form the sand clock.
Practise inverting the sand clock a few times.
Start timing the sand clock as soon as the bottles are inverted again.
Finish timing the clock when all the sand has collected in the bottom bottle.

## What to notice:

- when the sand started to flow
- how the sand moved from one bottle to another
- the length of time the sand took to collect in the bottom bottle

## More to do:

Use different quantities of sand in the clock to measure the time.
Change the size of the hole by placing a small cardboard disc in the opening.
Substitute sugar, salt, or rice for the sand in the clock.
Try to make a sand clock that measures an exact time span.

GENERAL

SCIENCE AREA: *Life*

SCIENCE SKILLS: *Observing, classifying, predicting, experimenting*

## For your information...

Bones are organic material mixed with calcium and phosphorous compounds. They form the framework or the skeleton of vertebrates (animals with backbones). Some bones, such as those found in humans, are hard and stiff, while others are soft and flexible, such as those found in fish. In order to support the body, calcium and phosphorous compounds give bones their necessary hardness and strength. The bone's outer layer is a hard, fibrous membrane that blends into a spongy material inside. The marrow-filled center of the bone manufactures red and white blood cells. Blood vessels supply the bones with food and remove waste materials. Skeletal bones are joined by ligaments. Joints allow for different kinds of movement that are determined by the shape of the bone and the ligaments at the ends where they join.

A butcher shop is a good source for a large and varied supply of bones. Boil chicken or fish bones for at least 40 minutes and scrub clean with a brush and disinfectant soap. Rinse and dry the bones thoroughly before putting them into the bags. Chicken backs and necks are good for examining how bones fit together. Fish bones are small, thin, roundish, smooth and often pointed, and flexible. The backbone of a fish is excellent for observing how a fish swims because of the movable joints between the vertebrae.

## Materials needed for the bag...

- clean, dry chicken and fish bones
- magnifying glasses of different magnification

## Words to know...

| | | | |
|---|---|---|---|
| calcium | porous | backbone | flexible |
| regenerate | marrow | vertebrate | joint |
| blood | rigid | brittle | ligament |
| phosphorous compound | | | |

## Now try this...

- Have the butcher cut a large beef shank bone into cross sections of 5 cm so students can observe the changes in the internal structure at different points along the bone.
- Leave cleaned bones in foil in a conventional oven over several weeks so that they are heated each time the oven is used. Monitor and record physical changes in color, mass, etc.
- Place a bone in a paper bag in a microwave oven to see what physical changes take place. Compare results with those achieved in a conventional oven.

## Look for these...

*Looking at the Body* by David Suzuki. Stoddart, 1988.

*Biology for Every Kid: One Hundred and One Easy Experiments that Really Work* by Janice Vancleave. J. Wiley & Sons, 1990.

*An Early Start to Technology* by Roy Richards. A Macdonald Book, 1991.

*Eyewitness Books: Skeleton* by Steve Parker. Stoddart, 1988.

*Anancy and Mr. Dry-Bone* by Fiona French. Little, Brown and Company, 1991.

*The Bones Book and Skeleton* by Stephen Cumbaa. Somerville House Publishing, 1991.

*Skeletons! Skeletons! All about Bones* by Katy Hall. A Platt & Munk All Aboard Book, 1991.

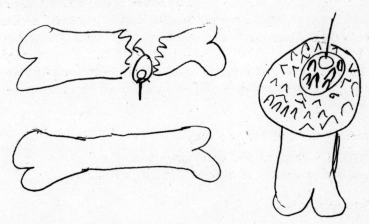

Break one chicken bone and look inside.

# Dry bones

**You will need:**

- clean dry bones from different parts of chicken and fish skeletons
- magnifying glasses of different magnification

**What to do:**

Look at each bone with the magnifiers.
Take the bones apart and examine the parts.
Break the bones to see the insides.

**What to notice:**

- how the bones are the same or different
- how the bones are joined
- what the inside of the bones looks like

**More to do:**

Try to put the bones back together again.
Think about how your bones are similar to these bones.
Soak chicken bones in vinegar for at least a week to see what happens.

# Overview — Take a breath

GENERAL

SCIENCE AREA: *Life*

SCIENCE SKILLS: *Observing, constructing, predicting, inferring*

## For your information...

This bag demonstrates how air pressure and the diaphragm work together to make the lungs work. When you breathe, air passes through the mouth and/or nasal passage, down through the trachea or wind pipe, and into the lungs. As the chest cavity expands, the lungs inhale air. When the chest cavity contracts, carbon dioxide, along with unused air, is exhaled. The diaphragm moves up, compressing the air within the chest cavity which increases the pressure and forces air out of the lungs. The diaphragm moves down, lowering the air pressure and drawing air into the lungs.

In the model in this activity, the stretched balloon on the bottom of the container moves up and down in a similar fashion to the diaphragm. The inside balloon represents the lungs and the straw represents the trachea.

## Materials needed for the bag...

- 2 L clear, plastic bottle
- straw
- 2 elastic bands (1 small, 1 large)
- Plasticine
- 2 balloons (1 small, 1 large)

## Words to know...

air pressure    diaphragm    inhale    exhale
trachea    lungs    chest cavity    oxygen
carbon dioxide

## Now try this...

- Reconstruct the model to include two lungs.
- Research respiratory diseases and the equipment used to help afflicted people breathe.
- Compare how other animals, such as insects, amphibians, reptiles, and fish breathe.

## Look for these...

*Biology for Every Kid: One Hundred and One Easy Experiments that Really Work* by Janice Vancleave. J. Wiley & Sons, 1990.
*Looking at the Body* by David Suzuki. Stoddart, 1988.
*how my body works: breathing* by Joan Gowenlock. Wayland, 1992.

# Take a breath

## You will need:

- 2 L clear, plastic bottle
- plastic drinking straw
- Plasticine
- scissors
- 2 balloons (1 small, 1 large)
- 2 elastic bands (1 small, 1 large)

## What to do:

Attach the small balloon to the straw with the small elastic band.
Cut the plastic bottle in the middle and keep the upper half with the neck.
Use the Plasticine to form a stopper around the middle of the straw.
Insert the balloon end of the straw into the neck of the bottle.
Make sure the stopper fits tightly in the bottle.
Cut the mouth end off the large balloon.
Attach the large balloon to the bottom of the bottle with the large elastic band.
Pull the large balloon down and push up gently.

## What to notice:

- what happens to the small balloon when the large balloon is pulled down and when it goes back up
- what happens to the plastic bottle

## More to do:

Try a different sized container.
Blow into the straw.

# Overview — Germinating seeds

GENERAL

SCIENCE AREA: *Life*

SCIENCE SKILLS: *Observing, constructing, predicting, measuring*

## For your information...

Seeds contain all that is needed to start new plants. The embryo is the miniature plant inside the seed. The seed stores food in the form of carbohydrates for the embryo to use. The seed coat is the protective layer surrounding the embryo and stored food. By soaking a bean seed in water overnight, removing the seed coat, and gently pulling the seed apart, the embryo can be removed and observed under a magnifier. When the new root pushes through the seed coat, sprouting or germination has begun.

In order for germination to occur, certain moisture, air, and temperature conditions must be met. In this activity, students provide seeds with the right conditions to allow them to germinate. The damp paper towel provides a constant source of moisture, while the plastic bag acts as a greenhouse by retaining the air at room temperature and maintaining the moisture level.

Vegetable seeds are generally the easiest to germinate, as well as being inexpensive and readily available in grocery or bulk-food stores. Packaged dried cooking beans are also suitable. Germination times will vary with the type of seed and the conditions present. A calendar is a good graphic organizer for tracking predictions and observations of germinating seeds. When more than one type of seed is used, consider color-coding calendar markings.

## Materials needed for the bag...

- small plastic bag
- paper towel
- seed, such as bean, pumpkin, or corn
- calendar
- pencil

## Words to know...

| | | | |
|---|---|---|---|
| seed | germinate | embryo | corn |
| pumpkin | sprout | carbohydrate | moisture |
| temperature | vegetable | seed coat | root |
| stem | bean | | |

**Now try this...**

- Research the two types of endosperm structure in seeds. Use an example of each type in a germinating activity.
- Determine the necessary conditions for germination by conducting experiments in which each variable is isolated in turn.
- Save, dry, and germinate seeds from fresh vegetables.

**Look for these...**

*How a Seed Grows* by Helene J. Jordan. Harper Collins, 1992.
*Up the Science Ladder: Activity Based Ideas for Teaching Science to Primary Grades* by Lynn Molyneux. Trellis Books Inc., 1988.
*Looking at Plants* by David Suzuki. Stoddart, 1988.
*Plants in Action* by Robin Kerrod. Marshall Cavendish, 1990.

# Germinating seeds

**You will need:**

- small plastic bag
- paper towel
- seed, such as bean, pumpkin, or corn
- calendar sheet
- pencil
- water

**What to do:**

Circle the date on the calendar when this activity is started.
Predict when the seed will germinate and mark the date on the calendar.
Moisten the paper towel with water.
Place the damp paper towel in the small plastic bag.
Put the seed on the paper towel and seal the bag.
Allow some air to flow into the bag (if not the seed will rot).
Set the bag aside in a sunny place where it will not be disturbed.
Check the seed each day and keep the paper towel moist.
Circle the date when the seed germinates and compare it to the prediction.

**What to notice:**

- how long the seed took to germinate
- the accuracy of the prediction
- which part of the seed started to grow first
- how the seed changed as it germinated

**More to do:**

Compare germination rates by using two or three different seeds and color-coding the information on the calendar.
Do not moisten the paper towel.
Place the bag in a dark place, such as a closet.
Plant the germinated seeds in small peat pots or egg shells.

GENERAL

SCIENCE AREA: *Life/technology*

SCIENCE SKILLS: *Observing, constructing, experimenting, predicting*

## For your information...

Decay or decomposition occurs when matter breaks down by natural or chemical processes. Some materials take longer to decompose than others. The rate of decay depends upon many factors. Under proper air, water, and temperature conditions bacteria in soil carry out decomposition. Generally, dead plants and animals decay in soil; inorganic matter, such as glass and metal, do not.

In this activity, which simulates a composter, students observe which materials decompose in soil. The organic objects are pieces of fruit or vegetable peel, newspaper, and styrofoam. The inorganic objects are a marble and a metal bottle cap or tab. Air is present in the soil and in the plastic bag, even though it is sealed. There is a certain amount of moisture naturally occurring in the soil, which can easily be maintained by adding a few drops of water from time to time. When this is done, air also enters the bag. The bag is kept at room temperature to allow soil bacteria to function.

This activity's time frame is open-ended. The fruit or vegetable peel generally decomposes first, followed by the newspaper within a month. The styrofoam, metal, and glass do not decompose, even after many months. It is suggested that weekly observations be made; a time line is a concrete way of recording changes over time. Discuss the fact that there is no time limit for this activity. When students realize that some materials do not decompose, they gain a sense of the impact these materials can have on the environment and the value of composting. Students then have the opportunity to use their experience from the activity for redesigning their own improved, mini-composters.

## Materials needed for the bag...

- clear plastic bag
- soil
- small piece of styrofoam
- pencil

- small piece of newspaper
- metal cap or tab
- time line

## Words to know...

| | | | |
|---|---|---|---|
| decay | decompose | compost | organic |
| inorganic | carbon | bacteria | moisture |

## Now try this...

- Design and build a composter for the school.
- Initiate a composting program in the school.
- Research local recycling programs.
- Investigate vermiculture for a year-round classroom project.

## Look for these...

*National Worm Day* by James Stevenson. Greenwillow, 1990.

*My First Green Book* by Angela Wilkes. Stoddart, 1991.

*Atlas of Environmental Issues* by Nick Middleton. Facts on File, 1989.

*How Green are You?* by David Bellamy. Clarkson Potter Publishers, 1991.

*Earthworms* by Robert Knott. Lawrence Science, 1989.

# Mini-composter

**You will need:**

- clear plastic bag half-filled with soil
- small piece of fruit or vegetable peel
- small piece of newspaper
- metal bottle cap or tab
- small piece of styrofoam
- pencil
- time line

**What to do:**

Lay the bag of soil on its side and flatten the soil evenly.

Press the peel along with paper, metal, and styrofoam objects into the soil.

Seal the bag and set it aside where it will not be disturbed.

Record weekly observations of the objects' appearance on the time line.

Add 20 to 25 mL of water to the bag from time to time.

**What to notice:**

- what kind of changes occur in each object over time
- how long the objects take to decompose
- what happens to the soil

**More to do:**

Repeat the activity with different inorganic objects.

Track and compare the decomposition of other organic objects.

Design a mini-composter using found materials.

# Overview — Colorful black

SCIENCE AREA: *Physical*

SCIENCE SKILLS: *Experimenting, inferring, hypothesizing, predicting*

## For your information...

Black contains the primary paint colors of red, yellow, and blue. When combined, primary colors absorb light and do not reflect it. The resulting color: black. One method of separating colors is by paper chromatography. Depending on the manufacturer's dye lots, black pens may have various mixtures of colors to make black ink. Chromatography helps you discover what colors were used to make the black ink in markers. Capillary action, the movement of liquids upward through small openings or spaces, is used in chromatography to separate the mixture of colors. In materials, such as paper filters, tiny connected spaces act as tubes, or capillaries, through which the water moves. As the water seeps through the paper, it carries the dyes with it.

In this bag, the liquid, a solution of black marker ink dissolved in the water, rises through the filter. The colors in the black ink separate at different rates. Generally, as the colors dissolve they are carried with the water to the edges of the filter paper. When drawing the bands or design on the filter, it is important to keep the center blank so the water being pulled up the paper is clean. It is always fascinating to see how black is a mixture of colors. If you have done any home decorating you can appreciate the importance of purchasing wall paper or getting all the paint mixed at the same time.

## Materials needed for the bag...
- coffee filter   • water-soluble marker   • clear plastic glass

## Words to know...

chromatography   pigment   primary colors   capillaries
capillary action   solution   dissolve   soluble

## Now try this...
- Use India ink sparingly in place of black markers.

- Research primary colors and color theory.
- Experiment with mixing colors using water colors.
- Tie dye t-shirts with students using natural vegetable dyes, such as onion skins for gold or beets for red.

**Look for these...**

*The Magic Fountain* by Alison Alexander. William Collins and Son, 1986.

*Rainbows to Lasers* by Kathryn Whyman. Gloucester Press, 1989.

*Light Fantastic* by Robin Kerrod. Marshall Cavendish, 1990.

*Science Demonstrations for the Elementary Classroom* by Dorothea Allen. Parker Publishers, 1988.

---

*Science Response Journal*
*Colorful Black*

*The experiment that our group worked on is sort of boring and sort of fun and surprising to me. The first time we did it we got it all wrong, we put the design right in the middle of the filter paper. The second time we got it right. The boring part of the experiment is waiting for 15 minutes, the fun and surprising parts are making designs near the middle of the filter paper and putting it in the water and seeing the designs and colors that come from the black marker.*

*My group tried some changes to the experiment. We tried different color marker but the marker that appears to have the most colors is the black marker.*

*Jonathan*

---

# Colorful black

**You will need:**
- flat, circular coffee filter
- water-soluble black marker
- clear plastic glass
- water

**What to do:**

Draw a design with the marker near the center of the coffee filter.
Fold the filter in half three times to make a cone.
Fill the bottom of the glass with water to about 1 cm.
Place the tip of the filter in the water for about 15 minutes.
Remove the filter, spread it out, and let it dry.

design with a black marker around the middle

Place filter in glass with water

colors have dispersed & separated

**What to notice:**
- where the water went
- how the black designs changed
- what colors appeared
- how the filter looked after it dried

**More to do:**

Try different black markers.
Use other colored markers and compare the results.
Leave the folded filter in the glass overnight to see what happens.

# Overview — Slippery surfaces

GENERAL

SCIENCE AREA: *Physical*

SCIENCE SKILLS: *Observing, measuring, inferring, controlling variables*

## For your information...

Friction is the force that resists the motion of one surface over another. Two surfaces rubbing together cause friction and give off heat. For example, when hands are rubbed together there is friction and heat is felt. If the surfaces are smooth, such as a brick sliding across a tiled floor, less friction is created and they slide easily over one another. If the surfaces are rough, such as a brick on a carpeted floor, more friction is created, making the movement more difficult and slowing the object down. The standard unit used to measure friction is Newtons (N), or the amount of energy needed to move the object by overcoming the friction. The standard unit used to measure mass is grams (g). A spring scale or force meter is used to measure the force of friction. The amount of friction varies with the mass; the greater the mass, the greater the force of friction. If a lubricant, such as water or oil, is added the friction is reduced. Although friction slows objects down, it is beneficial in a number of ways. For one thing, it improves the stability of cars on the road.

In this activity students are given the choice of using a brick or a heavy book as the object to pull with the spring scale. They are also asked to choose three different surfaces over which to pull the object. The distance of one metre (m) is kept constant over the three surfaces by marking it with tape. Since the surface is the only variable, variety in surface texture produces greater differences in the amount of friction. Students are asked to monitor or be aware of the readings on the spring scale.

When the force is first applied, the measurement will be higher because of the initial force needed to start the motion or overcome the resistance. When the object moves at a constant rate, the force is also constant. The scale returns to zero when the movement stops. The graphic organizer to keep track of the force measured may look like that on the next page.

# FORCE MEASURED IN NEWTONS (N) for *SLIPPERY SURFACES*

|  | Initial force | Force while in motion | Resting force |
|---|---|---|---|
| Surface #1 |  |  |  |
| Surface #2 |  |  |  |
| Surface #3 |  |  |  |

## Materials needed for the bag...

- spring scale
- masking tape
- string
- graphic organizer
- measuring tape

## Words to know...

friction    lubricant    spring scale    Newton    force

texture    surface    meter    resistance    mass

## Now try this...

- Lubricate the surface with different liquids, such as water or oil.
- Research safety considerations regarding friction in the home or on the highway.
- Find out how friction is counteracted and utilized in the school and community..
- Discover the difference between mass and weight.

## Look for these...

*Physics for Every Kid: One Hundred and One Experiments in Motion, Heat, Light, Machines, and Sound* by Janice Vancleave. J. Wiley & Sons, 1991.

*An Early Start to Technology* by Roy Richards. A Macdonald Book, 1991.

*Dropping in on Gravity* by Ed Catherall. Chrysalis Publications, 1988.

*Science Express: An Ontario Science Centre Book of Experiments.* Kids Can Press, 1991.

# Slippery surfaces

**You will need:**
- brick or heavy book
- spring scale
- various surfaces
- string
- measuring tape
- masking tape
- graphic organizer

**What to do:**

Tape start and finish lines about a meter apart on three surfaces, such as tile, carpet, or wood.
Use the string to tie the scale to the brick or book.
Slide the brick or book from start to finish over the first surface several times.
Read the number on the scale while pulling the brick or book in a steady motion.
Record the forces from the scale.
Repeat the steps on two other surfaces.
Record and compare the resulting forces.

**What to notice:**
- the force measured at the starting line, the finish line, and in between
- any sounds heard while moving the brick or book
- which surface required the most, least, and medium amounts of force
- how the brick or book feels after it has been moved

**More to do:**

Find the averages of trials with the same object on two different surfaces.
Wet the surfaces and measure the forces again.
Use different objects and compare the results.

# Overview — Detecting prints

GENERAL

SCIENCE AREA: *Physical*

SCIENCE SKILLS: *Observing, classifying, experimenting, communicating*

## For your information...

The pattern of slightly elevated ridges of the skin on fingertips of each person is unique and unchanging. These ridges are formed by the double row of peglike structures called papillae, which anchor the epidermis (outer skin) to the dermis (inner skin). Fingerprint patterns are classified as loops, arches, whorls, and accidentals (combinations of the first three patterns). Because fingerprints are unique to individuals, they have been used for identification purposes. This practice is known as dactylography.

The activity in this bag involves detection of fingerprints left on dark surfaces. After the fingers have been rubbed through the hair, a slight amount of oil is left on them. When the fingers are pressed on the black paper, the oil is transferred from the ridges to the paper. The resulting markings are generally not visible to the human eye. Therefore, the black paper is dusted with talcum powder so that it adheres to the markings and makes the fingerprints visible.

Forensic scientists have perfected and used print detection in solving crimes for many years. Examine your windows before you wash them to see if any prints can be detected.

## Materials needed for the bag...

- magnifying glass
- black construction paper
- talcum powder

## Words to know...

dactylography  arches  loops  whorls  accidentals
ridges  dermis  epidermis  papillae  talcum powder

## Now try this...

- Research the use of fingerprint identification in forensic science.
- Use an inked stamp pad to obtain fingerprints on white paper.

- Survey the fingerprint types of a group of students.
- Investigate fingerprinting practices at a local police station.

**Look for these...**

*Looking at the Body* by David Suzuki. Stoddart, 1988.
*The Magic School Bus Inside the Human Body* by Joanna Cole. Scholastic, 1988.
*Biology for Every Kid: One Hundred and One Easy Experiments that Really Work* by Janice Vancleave. J. Wiley & Sons, 1990.
*Fingerprinting* by Jeremy Ahouse. Lawrence Science, 1987.

# Detecting prints

**You will need:**

- magnifying glass
- talcum powder
- black construction paper

**What to do:**

Rub your fingers through your hair several times.
Gently place your fingers in the talcum powder.
Press all of your finger tips firmly on the black paper.
Use the magnifying glass to observe the line patterns made by the fingerprints.
Describe your observations.

**What to notice:**

- what patterns your fingerprints make
- whether the fingerprints are all the same
- how the fingerprints compare with one another

**More to do:**

Compare the fingerprints of other people with your fingerprints.
Make fingerprints without rubbing your hands through your hair.
Find out how fingerprints are classified.

# Overview — Speedy marbles

GENERAL

SCIENCE AREA: *Physical/technology*

SCIENCE SKILLS: *Observing, constructing, measuring, experimenting*

## For your information...

Moving objects either accelerate, decelerate, or move at a constant speed. The surface, the slope, and the object are variables affecting the rate of speed. For example, an object sliding or rolling down an inclined plane will accelerate. In order to slow or change the direction of the moving object another force must act upon it.

In this activity, an inclined plane is created by slanting the table. This allows the marble to be put into motion. When the marble is rolled for the initial timing, the table's smooth surface offers little resistance and the marble travels swiftly from one end to the other. The marble gains speed as it moves down the slope. This has to be taken into consideration when the track is being designed to slow the marble down. The newspaper track then operates as the external force to slow the marble's acceleration.

Designing the track is tricky and requires much thought. Trial and error is essential to meet the challenge presented in this bag and should be encouraged. If the angle of the newspaper track creates a barrier to its motion, the marble might come to a stop. In some cases, if the speed of the marble is too fast and the walls of the track are too low, the marble may jump over them. Therefore, the angle and the height of the newspaper track are important to direction, acceleration, and deceleration.

Another consideration is the overall distance. Setting a specific time of 15 S means that the marble must cover a certain amount of ground. This requires knowledge or at least an awareness that the distance travelled is equal to the rate of speed multiplied by the time.

You can experience this acceleration when travelling down a mountain road. Hairpin turns and switchbacks control the acceleration and the angle of descent. Look for this the next time you go on a water slide or a bobsled run. Students have the opportunity to incorporate variables when designing their own maze, and some may be challenged to invent a "Rube Goldberg" device.

**Materials needed for the bag...**
- marble
- 2 small blocks of wood or 2 books
- masking tape

**Words to know...**

| acceleration | deceleration | motion | inclined plane |
|---|---|---|---|
| distance | Newton | Rube Goldberg | |

**Now try this...**
- Vary the moving object and the type of surface used.
- Extend the same activity to one involving a "hot wheel" track and cars or a "Marbleworks" set.
- Research, design, and construct a Rube Goldberg machine.

**Look for these...**

*Raceways: Having Fun with Balls and Tracks* by Bernie Zubrowski. Morrow Junior Books, 1985.

*Physics for Every Kid: One Hundred and One Experiments in Motion, Heat, Light, Machines, and Sound* by Janice Vancleave. J. Wiley & Sons, 1991.

---

*Science Response Journal*
*Speedy Marble*

The experiment was hard because you had to make a track for the marbles. It looked like it was easy but when you tried to make the track it was very hard. We had to make it for 15 seconds.

The equipment we used was a marble, a long table, masking tape, news paper, two books and a stop watch or a second hand. It was hard to make what we wanted because to make the track we needed everybody to help but not everybody did so we got some people to help us. But after three days we were done.

I liked the fact that we had to make a track because it feels like we were making a subway. I don't like the way we had to make it because it was too long and because it took to much time.

When we were done the track we put the marble down it but it did not work so we fixed it and made tunnels. We used books to make slopes so the marbles could go down faster and it worked and so did the tunnels.

*Tristan*

---

# Speedy marbles

**You will need:**

- 1 marble
- long table
- masking tape
- newspapers
- 2 small blocks of wood or 2 books
- stop watch or watch with a second hand

**What to do:**

Place the books or blocks of wood at one end of the table under two legs so that the table is raised about 8-10 cm off the floor.

Use newspapers and tape to design a track to guide the marble.

Roll the marble along the track from the high end of the table to its low end.

Time how long it takes to go from one end of the table to the other.

Change the track design so that the marble takes 15 S to travel from one end to the other.

Draw the final design of the track and record how it was changed.

**What to notice:**

- what was done to slow the speed of the marble and why it worked
- what was done to increase the speed of the marble and why it worked
- what the final shape of the track was that accomplished the goal of 15 S.

**More to do:**

Repeat the activity by changing the slant of the table and compare the results.

Use a larger marble or a ball and compare the results.

Design a maze with several hills, switchbacks, and hairpin curves.

# 3

# Infusion strategies

In an integrated classroom program, themes provide a contextual framework to develop understanding and bring meaning to the teaching/learning process. Themes also provide a coherent experiential base for children's learning of distinct school subjects. Infusion is a means of incorporating new ideas, activities, and subjects into those themes. Teachers accomplish this by drawing upon appropriate infusion strategies, such as redrafting existing themes, designing new themes, and differentiating existing and new themes. The Sign Out Science Infusion Model demonstrates these processes.

Generally, themes are built by taking a kernel of an idea and expanding upon it in a sequential format. Those involved in the planning decide on a theme. In a brainstorming session they generate ideas and activities about the knowledge, attitudes, and skills to be covered. Information from this exercise is recorded under subject area headings on a graphic organizer. Webs are workable planning tools that help to organize activities and develop Sign Out Science bags. (See Chapter One for information on webs and the development process.)

There are many ways to integrate science into the curriculum. This chapter shows how three strategies infuse Sign Out Science

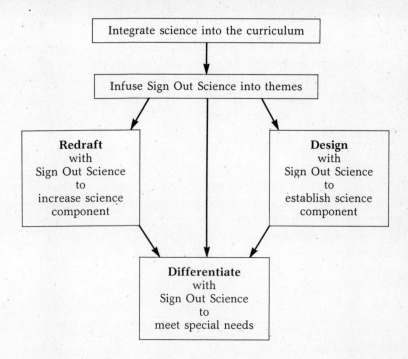

into themes. When the program is already in place, the quickest way to infuse Sign Out Science is to use an existing theme as a springboard. Original themes can be designed to fulfil new program requirements or go in new directions. As exceptional students are mainstreamed into regular classrooms, differentiating themes helps to meet their needs. Three popular themes are selected. In one, *Apples*, the science component is increased as the existing theme is redrafted. *Ice and snow*, a new theme, is designed to establish the science component. In the theme of *Motion*, the science component is differentiated to meet the needs of exceptional children.

## Redraft with Sign Out Science

Many teachers have already developed themes that fit into their curriculum and have worked successfully with students. These themes can be continued, but redrafted, so they have a stronger science component. The advantages of redrafting are numerous. Themes that have been used previously are more economical in terms of resources, funds, time, and energy. If the theme has been successful, then it is more comfortable and efficient to continue using it.

When redrafting an existing theme, the goal is to increase the science component. However, this may be a good time to make other modifications. Less successful or outdated activities can be replaced. New technology and other resources can be introduced. The purpose of some activities may change. Parental and community expertise can be tapped. The following web is the first example in tracking the process of redrafting the existing theme, *Apples*.

Begin by surveying the theme web for what is already there. Identify the type and number of activities listed under the various subject areas. Think about existing resources and materials. Examine the activities more broadly in terms of science as some may be modified to include a science focus. This requires shifting gears into a creative mode and becoming more aware of the science content in unusual places. Although all the activities may not lend themselves to science, ask questions, such as the following, to see if any aspect of the activity can be seen as scientific:

- Why is this happening?
- What makes it happen?
- Does it always happen in the same way?
- What would happen if I...?

The present *Apple* theme lacks sufficient science content. It may be useful to set a goal of a percentage of science activities in the theme. To reach the goal, activities will have to be developed and added. This can be done by either modifying an activity in another subject area or creating a new one. For example, one of the social studies activities is to research apple customs around the world. One custom is the well known bobbing for apples activity. Ask, "What makes the apples float?" and you have the creative spark for a science activity. The bag entitled *Bobbing apples*, on page 83, investigates how apples bobbing in water follow Archimedes' principle.

When creating a new activity, consider the learning objectives and the science concepts. For example, if listening for a purpose is one of the objectives, then the concept may be how sound is produced. The bag entitled *Seed rattles*, on page 88, investigates how seeds in containers produce sounds when shaken.

As a result of the redrafting process, four Sign Out Science bags, drawing from technology and the three areas of science, were added to the theme's science component. In language arts, a word-list activity was replaced with a computer wordsearch and a new writing activity was added. A community resource was brought in to augment the health content. A research component about pioneers was added to social studies. In visual arts, the apple print activity became more practical and a creative thinking activity was added. A graphing activity was included in math.

ORIGINAL THEME WEB

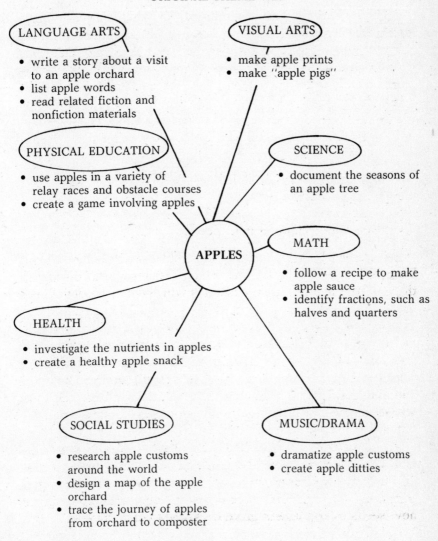

**LANGUAGE ARTS**
- write a story about a visit to an apple orchard
- list apple words
- read related fiction and nonfiction materials

**VISUAL ARTS**
- make apple prints
- make "apple pigs"

**PHYSICAL EDUCATION**
- use apples in a variety of relay races and obstacle courses
- create a game involving apples

**SCIENCE**
- document the seasons of an apple tree

**APPLES**

**MATH**
- follow a recipe to make apple sauce
- identify fractions, such as halves and quarters

**HEALTH**
- investigate the nutrients in apples
- create a healthy apple snack

**SOCIAL STUDIES**
- research apple customs around the world
- design a map of the apple orchard
- trace the journey of apples from orchard to composter

**MUSIC/DRAMA**
- dramatize apple customs
- create apple ditties

# REDRAFTED THEME WEB

**LANGUAGE ARTS**
- write a story about a visit to an apple orchard
- generate a computer wordsearch puzzle
- read related fiction and nonfiction materials
- create a new ending for *The Oldest Tree*

**VISUAL ARTS**
- design apple-print book covers
- make "apple pigs"
- make a model of an apple-picking device

**PHYSICAL EDUCATION**
- use apples in a variety of relay races and obstacle courses
- create a game involving apples

**SCIENCE**
- document seasons of an apple tree
- Bobbing apples (Sign Out Science)
- Seed rattles (Sign Out Science)
- Seedy fruits (Sign Out Science)
- Apple juicer (Sign Out Science)

**APPLES**

**HEALTH**
- investigate the nutrients in apples
- create a healthy apple snack
- invite a nutritionist to visit

**MATH**
- follow a recipe to make apple sauce
- identify fractions, such as halves and quarters
- graph the number of seeds found in ten apples

**SOCIAL STUDIES**
- research apple customs around the world
- design a map of the apple orchard
- trace the journey of apples from orchard to composter
- investigate the ways pioneers used apples

**MUSIC/DRAMA**
- create apple chants
- dramatize apple customs

# Design with Sign Out Science

As mentioned in the section on *Collaborative planning* in Chapter One, developing a theme involves integration of several subject areas and requires careful planning. It provides an opportunity to incorporate new approaches and new ways of looking at curriculum. In planning, it is useful to think across subject area boundaries, consider scientific skills, remember the scientific method, and accommodate students' learning styles.

Designing a new theme is an opportunity to use current ideas in a fresh way. Student and teacher interests, special events, or popular issues can contribute to the theme's development. Once into this process, serendipity comes into play. Up-to-date materials, resources, and expertise stimulate further ideas and contacts, which in turn open new directions.

New ideas for theme activities may be generated from a variety of sources. Completed charts on what the class knows and would like to find out can be used. Observations of children's prior knowledge, developmental stage, interests, and skill level determine the scope. Outside factors also need to be considered. Activities need to be in tune with policy and guidelines set by the educational system. As well, from time to time, broad societal expectations reflected in the media influence activity choices.

To facilitate the organization of a theme, the activities are generally grouped according to predominant subject areas; however, most activities contain aspects of other subject areas as well. For example, activities designated as science may also engage students in language arts, math, and music.

When thinking across subject area boundaries, teachers find organizational tools, such as a cross-curricular matrix, useful. The focus of this matrix is science activities and their overlap with other subject areas. This overlapping is important because in busy classrooms, crossing curriculum lines ensures that all subject areas are covered. Activities are rarely confined to one subject area, as in the Sign Out Science bag, *Acid snow*, on page 94. Although the predominant subject is science, language arts is included when students are asked to read and write, and math is included when students measure and graph.

After developing the web of learning activities, use a cross-curricular matrix, such as the one on page 75. The theme is recorded at the top and comments are placed at the bottom. The

Sign Out Science titles are listed down the side and subjects across the top. The resulting boxes provide places to record the presence of all subject areas and to check their distribution at a glance. For example, *Acid snow*, an Earth and Space science bag, covers language arts, social studies, and math.

The nature of the theme and interests may result in a matrix overloaded in certain subject areas. This can be balanced in several ways: change the focus of some of the science activities or address those areas that are left out when planning another theme. If the latter route is chosen, putting the matrices on transparencies will allow them to be used as overlays. The layers will then clearly show the subject area distribution and facilitate checking for balance.

Just as science activities overlap other subject areas, the science inquiry skills, as listed here, are the same as skills acquired in other subject areas and are important in the learning process. In most cases, the name of the skill is the same.

Science inquiry skills:

| *simple process skills* | *complex process skills* |
|---|---|
| classifying | experimenting |
| communicating | controlling variables |
| inferring | hypothesizing |
| measuring | interpreting |
| observing | making models |
| predicting | using equipment |
| seriating | |

Science inquiry skills, such as communicating and predicting, are common to language arts. Seriating and measuring are found in math. Observing and classifying are used for studying natural phenomena in science and human phenomena in social studies. Finally, observing and manipulating equipment are skills used for performing experiments in science and for aesthetic purposes in visual arts. The Sign Out Science bag, *Icy icicles*, on page 99, involves the science skills of communicating, experimenting, hypothesizing, and making a model. At the same time, students are using writing and speaking skills associated with language arts and measuring and estimating skills found in math. As students are engaged in what may appear to be strictly science activities, they are, in fact, simultaneously reinforcing skills learned in other parts of the curriculum.

# CROSS-CURRICULAR MATRIX

Theme title: _____  Subjects: _____

| Sign Out Science bags | Art | Language Arts | Mathematics | Music | Science | Social Studies |
|---|---|---|---|---|---|---|
| Earth and Space science<br>*Acid snow* | | | | | | |
| Life science<br>*Frozen fruits and veggies* | | | | | | |
| Physical science<br>*Icy icicles* | | | | | | |
| Technology<br>*Snow gauge* | | | | | | |

Comments: _____

Date: _____

_____

Some teachers may prefer having students use the process traditionally called the scientific method in which they ask specific questions, such as, "How are icicles formed?" and investigate to find the best answers. The scientific method is a way of organizing thought and action that has proved useful in learning more about the world around us. It takes the learner through a step-by-step, sequential inquiry process. It is the structure for controlling variables and replicating experiments.

---

*Steps in one model of the scientific method:*
1. State the hypothesis or ask the question.
2. Design an experiment to test the hypothesis or answer the question.
3. Determine the conditions or variables to be tested.
4. Determine what methods will be used to test the hypothesis.
5. Observe what happens for each condition or variable.
6. Analyze the observations.
7. Draw conclusions.
8. Evaluate results in light of original questions.

---

## Differentiate with Sign Out Science

Educators are responding to a major philosophical change in the education of students with special needs. For many exceptional students this means being integrated into regular classrooms. As a result, teachers will be expected to teach a much more heterogeneous group of students than before. A broader base of teaching skills and strategies will be needed. One strategy is to differentiate the curriculum to allow for individual differences within the classroom. Therefore the goal in this section is to investigate ways in which the science activities in the Sign Out Science bags can be modified to meet those students' varied needs.

Some differentiation is already inherent in the Sign Out Science bags. The bags are built upon inquiry and the questions "What, where, when, why, who, and how." Containing the necessary materials to conduct a highly motivating hands-on activity, the bags open with a simple, one-variable experiment. Clear and concise instructions guide students purposefully to employ all their senses and to notice as many details as possible. Extensions stimulate further investigation by suggesting activities with more complex experiments. Students can write or draw a response to what they discovered, and can read what others have experienced when doing the same bag. The bags allow for a wide variety of students with varying abilities. However, there are times when other modifications are necessary to ensure equal access of opportunity so exceptional students can actively participate and build self esteem.

Some teachers may be working in situations in which exceptional children are provided special help in a resource-room setting. If this is the case, the resource-room teacher might modify the Sign Out Science bags for students so they can participate in the science component of the theme. However, with exceptional students in the classroom, teachers have several alternatives to differentiate their theme and the Sign Out Science bags. Their chosen activities can reflect a wide variety of strategies and approaches, they may modify the bag in some way, or they may choose to do both. Many sensitive teachers already do many of the modifications listed below and on the next page, but these suggestions are offered as reminders when thinking about differentiating the bags for children with various exceptionalities.

With the exceptional student who takes longer to learn concepts, it makes sense to provide concrete experiences related to content that is meaningful. The bags do that by providing hands-on experience. For these students, teachers present tasks in an uncomplicated, brief, and sequential fashion — going from simple to complex. It is always important to go from the known to the unknown, especially with these learners. Since these students require repetition and overlearning to gain mastery, teachers need to have a variety of methods and materials at their command to teach the same concept. At the same time, it must be remembered that these students are capable of being taught metacognitive strategies, such as self-monitoring, in which they can talk themselves through a problem. They can also be shown explicitly how to apply their learning in other situations. Knowing this helps when thinking about the kind of modifications to make to the bags.

To modify Sign Out Science bags for students who take longer to learn concepts, teachers may wish to:

- rewrite the instructions in short simple sentences

- use drawings to show procedures
- rewrite extensions to reinforce the same concept
- have the student do the activity in class before she or he takes it home
- have the student repeat one of the variables several times
- explain where the experiment fits in to his or her everyday life
- make sure parents are available for guidance.

Students who have learning difficulties present a greater challenge because of their incredible range of characteristics and differences. However, research into learning style preferences provides some insights. Students experiencing learning difficulties are generally less persistent, less motivated, need more structure, and have more of a preference to learn from adults. They need a warm supportive environment and an understanding of their learning difficulty. With exceptional students, most teachers attempt to learn as much as possible about the student's unique characteristics. They understand that organization and structure can be crucial, and that consistent actions and expectations matter to these students. Based on this knowledge, teachers try to eliminate distractions, preface remarks with main ideas of the lessons to be taught, provide a focus on important points, check for immediate recall, and vary assignments so that both written and spoken feedback are included.

With some students, alternative testing formats and longer time lines are essential. Graphic organizers, such as webbing and semantic mapping, help to organize their ideas. Because many of these students don't have a broad repertoire of efficient learning strategies and do not generalize those few they may have, these skills and strategies must be taught explicitly. Training in metacognitive strategies provides them with a set of self-instructional steps to increase their effectiveness in acquiring, organizing, and expressing information. Many of these students, especially those who find writing difficult, respond well to computers because word processing enables them to manipulate text efficiently.

As well, students who are "learning different" need a sense of purpose and an understanding that learning involves developing products to solve real problems for real audiences. Therefore, they tend to be more motivated when given concrete, hands-on experiences with relevant outcomes that the science bags provide.

To modify Sign Out Science bags for students who are "learning different," teachers may wish to:

- rewrite and illustrate bag instructions
- number all instructions
- allow short written responses accompanied by drawings
- allow written and graphic responses on computer
- use as many visuals as possible
- include graphic organizers to record results
- incorporate a tape recorder for verbal responses
- encourage parents to facilitate students' work on the bags
- word extensions carefully to make room for growth.

Students who require more stimulating educational experiences also present challenges to teachers. Often having insatiable curiosity, these students are inquisitive, committed to finding out why things happen, able to think abstractly, and verbally fluent. In spite of these positive attributes, they also have special needs and should not be left to their own devices to learn. Differentiating for them can emphasize cognitive processing, abstract thinking and reasoning, creative problem-solving, and self-monitoring. Research into their learning styles indicates they generally tend to be motivated, persistent, and prefer to learn alone in a quiet environment. Seen as independent and nonconforming, they can be intimidating; however, many teachers already understand the techniques that enhance motivation and provide for their personal development.

Enrichment for these students is essential, but it must involve more than tacking on a few activities to keep them busy. Although there are a number of successful enrichment models, Renzulli's *Enrichment Triad Model* is useful here as an example when thinking through their programming. It delineates three types of enrichment activities — the first two are also very appropriate for all students. Type I activities expose students to exciting topics, ideas, and fields of knowledge. Type II activities are designed to develop thinking processes, research and reference skills, and personal and social skills. Type III activities involve exceptional students in investigating real problems and topics using appropriate methods of inquiry. Students take on the responsibility for initiating, conceptualizing, and planning the project activity. This level of enrichment is rigorous, challenging, and highly motivating.

Although it is possible to present more advanced material to these students, it cannot be assumed that they have the basic skills. Teach the basic structure or principles of a discipline and then let them approach it as a specialist does. Introduce important ideas as early as possible, model metacognitive strategies by which they can discover and synthesize knowledge for themselves, and stress the processes of problem solving, as well as the final product. Written expression can be used for a variety of purposes, and they are expected to use a variety of different kinds of technology to produce their results.

To modify Sign Out Science bags for students who need challenging educational experiences, teachers may wish to:

- rewrite instructions to reflect a sophisticated vocabulary
- increase the complexity of the bags' contents, such as adding variables
- have students establish variables
- request hypothesis statements before they begin
- include a research and design component for further investigation
- encourage students to research and develop theme bags for others.

# 4

# Theme-related activities in bags

The previous chapter outlined the integration of Sign Out Science into the curriculum. In this chapter, Sign Out Science bags are provided for class use and to serve as models.

The *Overview* indicates the theme: Apples, Ice and snow, or Motion. The bag's title, science area, and main skills are listed. The *For your information* section introduces general concepts, explains the activity, and offers practical tips. Under *Materials needed for the bag* is a list of suggested items to include. Feel free to substitute items if they are not readily available and other appropriate items are at hand. Selected vocabulary is presented in *Words to know*; activity extensions for teachers are suggested under *Now try this*; and related fiction and nonfiction material appear in *Look for these*.

The instruction sheet intended for students and parents begins with a list of materials found under *You will need* — most materials need to be included in the bags while others, such as water, scissors, or glass jars, are readily available at home. *What to do* provides step-by-step instructions, and is followed by guided observations in *What to notice*, and extensions in *More to do*.

Although technology is inherent in each of the three science areas — Earth and Space science, Life science, and Physical science — it has been accorded extra recognition, in order to heighten awareness and understanding of its presence and value. Technology involves drawing upon past experiences and applying that knowledge to a design technology where the end product may be the creation of prototypes and systems. In order to create, change, and improve these prototypes/systems, students examine the materials, the design, and the use of scientific inquiry, convergent, divergent, and critical thinking. Each set of bags is categorized by theme and science area. The first three bags in each set are labelled with the theme explored, while the fourth bag also contains a technology focus.

Four main science areas pertaining to the bags are drawn from the recognized science inquiry skills appropriate for elementary students. They include: classifying, communicating, controlling variables, experimenting, hypothesizing, inferring, interpreting, making models, measuring, observing, predicting, and seriating. Although these skills are listed separately, they do not exist in isolation. Instead, they are dependent upon the science content at hand, as well as the purpose and context of students' inquiry. Scientific inquiry is holistic and incorporates critical thinking in an integrated approach that enables students to derive meaning and understanding from their explorations.

# Overview — Bobbing apples

THEME: *Apples*

SCIENCE AREA: *Earth and space*

SCIENCE SKILLS: *Experimenting, communicating, interpreting,*

*measuring*

## For your information...

Archimedes, a Greek scientist, discovered that when an object is put into water and it floats, it displaces the water and there is an increase in the water level. Conversely, the water level decreases and returns to its original level when the object is removed. The larger the object, the more water it displaces, and the higher the water level rises. The amount of displaced water is equal to the mass of the object.

Apples bobbing in water is an example of floating displacement. The water that is displaced in the container is the same as or equal to the apples' mass. For practical purposes, 1 mL of water equals approximately 1 g, thus if the mass of the apples is 200 g the amount of water that is displaced would be 200 mL. For this bag, place a sticky label marked with the words *water line* horizontally near the top of the clear, plastic container. Place a sticky label perpendicular to the water-line tape. Students are asked to record their observations in a series of three pictures.

Be sure to notice how the water rises in the bathtub the next time you take a bath. The displaced water is equal to the mass of your body.

Step # 1

The water coms up to the water line.

## Step # 2

I think the water will rise.

I think the water will rise becaus when you put the appel in it tacks up rume. So than the water has to move up.

## Step #3

I tuck the appel out of the water. It went back to the same.

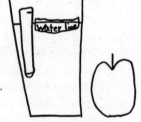

**Materials needed for the bag...**
- plastic knife
- plastic container, with a line marked at least halfway up the container
- pencil
- paper

**Words to know...**

displacement      volume      mass      Archimedes

**Now try this...**
- Determine the volume of water that an apple displaces by using a calibrated container.
- Determine the mass of water that an apple displaces by placing the water level to the top of the container, collecting the overflow, and weighing it.
- Research Archimedes' principle.

**Look for these...**

*The Red Apples* by Marion Mineau. Black Moss Press, 1986.

*Apples: All about Them* by Alvin Silverstein and Virginia B. Silverstein. Prentice-Hall Inc., 1976.

*Who Sank the Boat?* by Pamela Allen. Hamish Hamilton, 1982.

*Mr. Archimedes' Bath* by Pamela Allen. Angus and Robertson, 1991.

# Bobbing apples

**You will need:**

- small apple
- plastic knife
- labelled container
- pencil
- sheet of paper

**What to do:**

Fill the container with water to the line and draw a picture of it.

Predict what will happen to the water level when the apple is put in.

Lower the apple into the water, mark the new water level on the vertical label, and draw another picture.

Remove the apple from the water, observe the water level, and draw a third picture

**What to notice:**

- what happened when the apple was placed in the water
- what happened when the apple was taken out of the water

**More to do:**

Repeat the activity with one-half of the apple.

Repeat the activity with one-quarter of the apple.

Compare these observations with the original ones.

# Overview — Seedy fruits

THEME: *Apples*

SCIENCE AREA: *Life*

SCIENCE SKILLS: *Classifying, communicating, observing, interpreting*

## For your information...

After fertilization, many changes occur when blossoms turn into fruit. Petals fall off, seeds develop, and ovaries swell. Apples belong to a family of flowering plants whose parts grow in multiples of five. Their cores are characterized by a five-pointed, star shape comprising five compartments or carpels. The inner part of the ovary, the core, has a thin layer that wraps around the seeds to protect them. The outer layer, the receptacle, becomes the eating part of the apple. Because there are two ovules in each of the five carpels, a maximum of ten seeds can be produced. The ripened ovary of plants that contain seeds are known as fruit. Botanically, the term *pome* refers to apples, pears, and quinces, which all have a fleshy layer and papery core surrounding the seeds.

In this activity students discover an apple core's configuration. Suggestions for the second piece of fruit are drawn from the same family as apples to highlight the similarities. Suggestions for the third piece of fruit are drawn from other families whose seeds and cores are unlike those of apples to highlight the differences.

## Materials needed for the bag...

- magnifier
- plastic knife
- pencil
- paper

## Words to know...

carpel    fruit    core    pome    ovules    ovary

## Now try this...

- Find out how seed development affects an apple's shape.
- Graph the number of seeds in several apples to determine the minimum and maximum number of seeds.
- Make star-shaped apple prints using the cut apples.
- Make baked apples in class.

## Look for these...

*The Apple and Other Fruits* by Millicent E. Selsam. William Morrow and Company, 1973.

*The Oldest Tree* by Kathryn Guthrie. Reed Methuen Publishers Ltd., 1986.

*Projects for Fall* by Celia McInnes. Wayland, 1988.

*Exploring Nature around the Year: Fall* by David Webster. Julian Messner, 1989.

*Get Growing! Exciting Plant Projects for Kids* by Lois Walker. Pembroke Publishers Ltd., 1990.

## Seedy fruits

**You will need:**

- magnifier
- apple
- pear or quince
- orange, banana, or plum
- plastic knife
- pencil
- sheet of paper

**What to do:**

Cut each fruit across the middle between the stem and blossom ends.

Look at the seeds and what surrounds them.

Compare the apple seed section with that of the other two fruits.

Draw what you see for each fruit.

**What to notice:**

- how the seeds in each fruit are arranged
- which fruit is like the apple's seed section
- which fruit is unlike the apple's seed section

**More to do:**

Compare the seed sections of other fruits.

Discover other members belonging to the same category as apples.

Research how the formation of seeds and seed sections are related.

# Overview — Seed rattles

THEME: *Apples*

SCIENCE AREA: *Physical*

SCIENCE SKILLS: *Communicating, controlling variables, hypothesizing, observing*

## For your information...

Vibrating objects may produce sound. Sounds can be differentiated on the basis of loudness or pitch. Loudness increases when vibrations of sound-making objects have greater amplitude or when vibrating objects are closer to the ear. Rattles are percussion instruments in which objects inside a container strike its walls. Most percussion instruments produce musical notes that have only one pitch. A variety of sounds can be produced depending on the container and the objects inside. In this activity, small plastic film containers are used, leaving the seeds as the variable to be studied. Generally, dense and stiff objects bounce better and absorb less of the impact. This creates short, sharp sounds. Objects that are less dense or are soft don't bounce as well and absorb the impact. This creates more of a rustling sound.

In these rattles, ripened apple seeds have hard, smooth, rounded surfaces that contrast to the sound made by orange and pumpkin seeds that are more irregular in shape and have rougher surfaces. This activity helps students discover the differences in sounds, as well as the objects that create the sounds. Although color is a variable, it does not affect the sound. A graphic organizer to record observations might look like the following chart:

### Observations for Seed rattles

|  | sound | size | color | shape | feel |
|---|---|---|---|---|---|
| apple seeds |  |  |  |  |  |
| orange seeds |  |  |  |  |  |
| pumpkin seeds |  |  |  |  |  |

Many cultures use natural materials to create percussion instruments, such as maracas and shak-shaks made from gourds.

## Materials needed for the bag...

- 3 numbered containers (e.g., 35 mm film containers)
- apple, orange, and pumpkin seeds
- graphic organizer
- pencil
- magnifier

## Words to know...

amplitude     percussion     frequency     maraca
fundamental     pitch     vibrations     sound
shak shak

## Now try this...

- Place the three kinds of seeds in containers which differ, either in size or composition, from those containers used in the bag.
- Change the variable by using apple seeds in three different containers.
- Use immature seeds from unripe apples to obtain a different quality of sound.

## Look for these...

*An Apple Tree throughout the Year* by Claudia Schnieper. Carolrhoda Books Inc., 1987.
*Sound Science* by Etta Kaner. Kids Can Press, 1991.
*Science Fun with Drums, Bells and Whistles* by Rose Wyler. Julian Messner, 1987.
*Exploring Nature around the Year: Fall* by David Webster. Julian Messner, 1989.

# Seed rattles

**You will need:**

- 3 numbered containers with apple, orange, and pumpkin seeds
- magnifier
- graphic organizer
- pencil

**What to do:**

Shake each container in turn.
Listen to the sound each rattle makes.
Guess the type of seeds in each container.
Open the containers to check your guesses.
Spread the seeds out and look closely at them with the magnifier.
Complete the graphic organizer.

**What to notice:**

- how the rattles sound when different seeds are used
- how the seeds determine the sound each rattle makes
- which physical characteristic(s) affect the sound

**More to do:**

Repeat the activity using a variety of different seeds.
Repeat the activity using different containers.
Design an experiment to prove which physical characteristic(s) do(es) not determine the sound of the rattle.

# Overview — Apple juicer

THEME: *Apples*

SCIENCE AREA: *Physical/technology*

SCIENCE SKILLS: *Communicating, making a model, manipulating equipment, observing*

## For your information...

Technology includes the application of force to alter the appearance of raw materials in order to create products. Force can be applied by hand; however, machines create a greater force to save labor. Devices such as hand tools are useful because they extend our limited abilities and satisfy a need. Machines enable us to maximize our efforts.

In this activity, students use simple machines to extract juice from an apple. Students take the whole fruit through a series of changes by cutting, grating, and pressing the apple to extract the juice. The remaining pulp, seed, skin, and core are called pomace. These steps imitate the commercial process of producing both apple juice and cider. In a processing plant, apple juice is filtered and pasteurized or sterilized before being packaged. Students, using information gained by performing the process, have an opportunity to apply different materials to that process when they design their own juicers.

## Materials needed for the bag...

- small hand grater
- plastic knife
- wooden craft stick
- large piece of waxed paper
- small plastic container
- 30 cm square piece of cheesecloth

## Words to know...

cider    pasteurize    ferment    pomace    filter    sterilize

## Now try this...

- Repeat the activity by eliminating one aspect of the technology, such as the knife, the grater, or the stick.
- Clarify the juice by passing it through a coffee filter to extract remaining particles.

- Compare the total mass of the juice and pomace with the original mass of the apple and account for any discrepancy.
- Research tools used by pioneers.

**Look for these...**

*The Seasons of Arnold's Apple Tree* by Gail Gibbons. Harcourt Brace Jovanovich, 1984.

*An Apple a Day: From Orchard to You* by Dorothy Hinshaw Patent. Cobblehill Books/Dutton, 1990.

*An Early Start to Science* by Roy Richards. A Macdondald Book, 1987.

*Exploring Nature around the Year: Fall* by David Webster. Julian Messner, 1989.

# Apple juicer

**You will need:**

- apple
- small hand grater
- plastic knife
- wooden craft stick
- small plastic container
- 30 cm square piece of cheesecloth
- large piece of waxed paper

**What to do:**

Cut the apple in half.
Place the cheesecloth on waxed paper.
Grate each half of the apple over the cheesecloth.
Gather and knot the corners of the cheesecloth.
Pass the craft stick through two opposite openings under the knot.
Twist the stick and cheesecloth to squeeze the grated apple.
Collect the juice in the plastic container.

**What to notice:**

- how the grater changes the look and feel of the apple
- what process is needed to change the apple into juice
- how much juice is produced from the apple
- how the squeezed and grated apple pulp looks

**More to do:**

Design another way to get juice from an apple.
Investigate how cider presses operate.
Find an ecologically sound way to use the pulp.

# Overview — Acid snow

THEME: *Ice and snow*

SCIENCE AREA: *Earth and space*

SCIENCE SKILLS: *Communicating, observing, measuring, experimenting*

## For your information...

Acidity in rain or snow is caused by sulphur dioxide and nitrogen oxides emissions from combustion. The main sources of sulphur dioxide and nitrogen oxides are from coal-burning power generating stations and automobiles. These pollutants are released into the atmosphere and combine with water vapor in the clouds to form dilute acids. They are carried in the air by prevailing winds and come back as acid rain, snow, fog, or dust. The longer the pollutants remain in the atmosphere, the more they react with moisture to form an acid. Soil may contain materials or chemicals that neutralize the acids. When there is more acid than the soil can handle, it becomes acidic and affects the environment. This imbalance damages soil, vegetation, rivers, and lakes.

The experiment in this bag looks at acid precipitation. Acid levels in soil, snow, and rain can be measured. The pH scale measures the acidity and alkalinity. The scale goes from 0-14. The lower the number on the scale, the greater the acidity. On a pH scale, 7 is considered neutral, lemon juice at 2.3 on the scale is an acid, while milk of magnesia is a base at 10.5 on the pH scale. When using pH paper to test, a range of colors from reds to blues are used to indicate the acidity or alkalinity. Bright red indicates an acid and yellow-green indicates a base. In spring, the acid content in lakes and rivers is the highest because the acid, built up in the snow during the winter, is melting and producing a sudden rise in the acidity level of the water.

## Materials needed for the bag...

- 35 mm film containers
- labels
- pH paper
- swimming pool pH tester

## Words to know...

pH level   acid   sulphur dioxide   emissions   combustion
indicator   base   nitrogen dioxide   pollutants   alkaline

**Now try this...**

- Freeze a solution of vinegar and water, and crush or shave the resulting ice to make "snow." Compare the pH levels of the solution and the snow.
- Use other acidic liquids, such as orange juice or lemonade, and conduct the same experiment.
- Water one plant with melted snow and the second with tap water. Compare the two plants.

**Look for these...**

*Acid Rain* by Colin Hocking, Jacqueline Barber, and Jan Coonrod. Lawrence Science, 1990.

*My First Green Book* by Angela Wilkes. Stoddart, 1991.

*Science Express: An Ontario Science Centre Book of Experiments.* Kids Can Press, 1991.

*A New True Book Experiments with Water* by Ray Broekel. Children's Press, 1988.

*Atlas of Environmental Issues* by Nick Middleton. Facts on File, 1989.

*Exploring Nature around the Year: Winter* by David Webster. Julian Messner, 1989.

# Acid snow

**You will need:**

- snow or water from different outdoor locations
- 35 mm film containers
- labels
- pH paper
- swimming pool pH tester

**What to do:**

Gather snow and/or water samples from at least five different areas and place them in film containers.
Label containers with the name of each location.
Test for acidity following the instructions on both the pH paper package and the pH tester.
Graph the level of the acidity measured for each sample.

**What to notice:**

- how the pH levels are similar or different in the various samples
- how location and age of the sample affect the pH level

**More to do:**

Research the causes of acidity in rain and snow.
Find and test the acidity of icicles from different locations.
Make ice cubes using baking soda by adding two tablespoons of baking soda to a liter of water. Test the pH level. Compare the results with ice cubes containing lemon juice or vinegar.

# Overview — Frozen fruit and veggies

THEME: *Ice and snow*

SCIENCE AREA: *Life*

SCIENCE SKILLS: *Inferring, observing, experimenting, hypothesizing*

## For your information...

Vegetables and fruit are made up of cells that contain varying amounts of water. The cell walls consist of cellulose, which is rigid and holds the shape of the plant; however, freezing changes the texture. For example, when a crisp slice of frozen green pepper thaws it becomes limp and mushy. This change takes place as it freezes. Ice crystals that expand within the cells exert pressure and break the cell walls. As long as the vegetable or fruit is frozen, it will hold its shape. When it thaws, the broken cell walls collapse, causing them to lose their shape. Frozen beets placed in hot water demonstrate how cell membranes, broken during the freezing process, release color into the water.

This bag shows what happens when fruits and vegetables freeze and what they look and feel like when they thaw. Some fruits and vegetables, such as tomatoes and green peppers, change significantly when frozen and thawed.

## Materials needed for the bag...

- paring knife
- magnifying glass
- plastic bags

## Words to know

| | | | |
|---|---|---|---|
| cells | cellulose | ice crystals | pressure |
| expand | membrane | texture | |

## Now try this...

- Design a device to remove the water from frozen vegetables.
- Research *cryogenics* and *freeze dried*.
- List fruit or vegetables that contain large/small amounts of water.
- Describe the physical properties of fruit and vegetables listed in the above activity.
- Try freezing dried fruit and vegetables.

**Look for these...**

*175 Science Experiments to Amuse and Amaze your Friends* by Barbara Walpole. Random House, 1988.

*Exploring Nature around the Year: Winter* by David Webster. Julian Messner, 1989.

*The Pumpkin Blanket* by Deborah Zagwyn. Fitzhenry & Whiteside, 1990.

# Frozen fruit and veggies

**You will need:**

- fresh vegetables and fruit, such as lettuce, potato, grapefruit, and banana
- freezer or freezer compartment of a refrigerator
- paring knife
- plastic bags
- magnifying glass

**What to do:**

Put a slice of each fruit or vegetable in a bag.
Leave the bag in the freezer until it is frozen.
Record how long it takes for each item to freeze.
Observe each item under a magnifying glass.

**What to notice:**

- how juicy the fruit and vegetables are before/after freezing
- what the fresh/frozen/thawed items look like under a magnifier
- how the fresh/thawed items look and feel

**More to do:**

Conduct the same experiment with leaves from plants.
Find out which frozen vegetables and fruit are available at the supermarket.
Discover the effects of freezing on dried fruit or vegetables.

# Overview — Icy icicles

THEME: *Ice and snow*

SCIENCE AREA: *Physical*

SCIENCE SKILLS: *Communicating, experimenting, hypothesizing, making a model*

**For your information...**

The experiment in this bag explores the properties of water. Icicles are water in its solid state. Water, one of the most common substances, freezes and melts at approximately 0°C. Icicles are formed when water freezes and expands, and are usually found on the sunny side of a building because the sun's rays melt snow or ice. The melted water is pulled down by gravity, refreezes, and forms an icicle in an inverted cone shape.

The formation of icicles depends upon the rate of water flow and air temperature. In this experiment, the size of the hole determines the rate the water flows out of the bottle. If the hole is too big, it will form only a steady stream of water. If the hole is too small, the freezing water will plug the hole. The students should be cautioned to use only a few drops of food coloring because too much will lower the freezing point of water.

**Materials needed for the bag...**
- clear plastic bottles
- large needle
- tape
- food coloring
- string

**Words to know...**

freeze    melt    icicles    expand    substance

**Now try this...**
- Use one bottle of hot water and one bottle of cold water to time the formation of the two icicles.
- Add several drops of liquid detergent to the water and observe what happens.
- Bring icicles indoors to time how long they take to melt.
- Make edible icicles using liquids, such as different fruit juices.
- Research heat transfer and surface tension.

**Look for these...**

*Sadie and the Snowman* by Allen Morgan. Kids Can Press, 1985.
*Projects for Winter* by Celia McInnes. Wayland, 1988.
*We Celebrate Winter* by Bobbie Kalman. Crabtree Publishing, 1986.
*Water at Work* by Barbara Taylor. Franklin Watts, 1991.

## Icy icicles

**You will need:**

- clear plastic bottles
- hot water
- large needle or an awl to make a hole
- food coloring
- string
- tape

**What to do:**

Make different-sized pinholes on the side of each bottle about one-third of the way up the side of the bottle.
Cover the holes with tape.
Fill two plastic bottles 3/4 full (add a few drops of food coloring).
Hang the bottles outside.
Remove the tape.

**What to notice:**

- how the water flows out of each bottle
- how the size of the hole affects the icicle

**More to do:**

Redo the experiment filling one bottle with hot water and the other with cold water.
Look for icicles outside. Notice on which side of the buildings they are found. Observe their shapes and sizes.
Record and discuss your findings.

# Overview — Snow gauge

THEME: *Ice and snow*

SCIENCE AREA: *Earth and space/technology*

SCIENCE SKILLS: *Experimenting, measuring, manipulating equipment,*
*interpreting*

## For your information...

Technology involves designing and making objects that improve the way tasks can be performed. Some objects can be simple measuring devices, such as the ones that are used by meteorologists to record amounts of precipitation.

In this bag, materials are included for the collection of snow. Transparent plastic containers suit the purpose because they are nonbreakable, and the contents are easily observed. Various calibrations are marked on a strip of masking tape down the side of the container. This allows for standard and nonstandard measurements of the snow and the melt water (melted snow). Nonstandard measurements are evenly spaced.

Wind patterns should be taken into account when placing the snow gauge so that actual precipitation, not drifted snow, is collected. Open, sheltered areas near a wind break are good locations. The equivalent of 10 cm of snow is approximately 1 cm of water. Weather reports often include the amount of snowfall. Have you ever wondered how they measured it? Students may extend their experience gained from the activity by considering the additional factors of weather conditions, placement, and materials in order to design an improved snow gauge.

## Materials needed for the bag...
- clear plastic bottle
- masking tape
- permanent marker
- scissors

## Words to know...
| | | | |
|---|---|---|---|
| gauge | calibrate | sheltered | technology |
| design | weather | meteorologist | precipitation |

## Now try this...
- Use computer weather programs to design precipitation charts.

- Graph the levels of snow and melt water.
- Make prototypes of the gauges and determine the most effective ones.
- Write technical descriptions of the gauges.
- Develop a blueprint or design of the gauges that can be used by others.

**Look for these...**

*Weatherwatch* by Valerie Wyatt. Kids Can Press, 1989.

*Snowed In at Pokewood Public School* by John Bianchi. Bungalo Books, 1991.

*175 More Science Experiments to Amuse and Amaze Your Friends* by Terry Cash, Steve Parker and Barbara Taylor. Random House, 1990.

*Snow Company* by Marc Harshman. Cobblehill Books/Dutton, 1990.

*Tiger with Wings: The Great Horned Owl* by Barbara Juster Esbensen. Orchard Books, 1991.

*Girl from the Snow Country* by Masako Hidaka. Kane/Miller Book Publishers, 1986.

# Snow gauge

**You will need:**

- clear plastic bottle
- masking tape
- permanent marker
- scissors

**What to do:**

Cut off the top of the bottle.
Place a strip of masking tape down the side of the bottle.
Mark the calibrations to be used on the tape.
Find an open area to place the snow gauge.
Cement the gauge in place by packing snow and pouring some water around it.

**What to notice:**

- how long the snowfall lasted
- the level of snow the gauge collected after a snowfall
- the time it took to fill a calibrated unit

**More to do:**

Measure the melted snow and find the ratio of snow to melt water.
Investigate whether the same level of snow always produces the same level of melt water.
Make improvements on this design of a snow gauge.

# Overview — Around and around it goes!

THEME: *Motion*

SCIENCE AREA: *Earth and space*

SCIENCE SKILLS: *Observing, inferring, communicating, experimenting*

## For your information...

Centrifugal force pushes objects in an outward direction away from the center of a revolving body. As the earth spins around the sun, the force of gravity keeps it from flying off into space. Gravitational pull is the force that tends to draw objects toward the center.

In this activity, the balloon is moved in a circular motion making the object spin along the walls of the balloon. The spinning object inside the balloon tends to move away, but the balloon's walls keep it from flying off in a straight line. As a result, the object rotates in a circle and gains stability. As long as the object's rotation within the balloon's wall continues, it will remain upright and will not fall over. When the rotation slows, its stability is destroyed and the force of gravity takes over, causing the object to fall.

This same concept applies when drying lettuce in a salad spinner. The water spins away from the center, runs down the sides of the bowl, and is collected at the bottom of the container.

## Materials needed for the bag...

- 2 small light-colored balloons
- coin
- marble

## Words to know...

| | | | |
|---|---|---|---|
| stability | rotation | axis | gravitational pull |
| friction | counteract | exert | centrifugal force |

## Now try this...

- Research amusement park rides, industry, or medicine in which centrifugal force occurs.
- Experiment with a variety of loops in a track using hot wheels.
- Discover how these forces apply in nature, such as water going down a drain.

**Look for these...**

*A Balloon Goes Up* by Nigel Gray. Orchard Books, 1988.
*Balloon Building and Experimenting with Inflatable Toys* by Bernie Zubrowski. Morrow Junior Books, 1990.
*The SuperScience Discovery Book* by Professor Kurius. Grosvenor House Press, 1984.

---

# Around and around it goes!

**You will need:**

- 2 round, light-colored balloons • marble • coin

**What to do:**

Put the coin in one balloon, the marble in the other.
Blow up both balloons and tie the ends.
Hold the balloon containing the coin at both ends in front of you.
Quickly move the balloon in a circular motion away from you to start the coin spinning.
Hold the balloon still; watch and listen to the coin.
Quickly move the balloon in a circular motion toward you once.
Repeat the steps with the balloon and marble.

**What to notice:**

- how the coin and the marble react
- how the coin and marble move
- which moves faster, the coin or the marble

**More to do:**

Place other small objects inside balloons to see if they react in the same way.
Count the number of times, or rotations, an object goes around in the balloon.
Discover a way to increase or decrease the number of rotations.

# Overview — Pinch and lift!

THEME: *Motion*

SCIENCE AREA: *Life*

SCIENCE SKILLS: *Observing, interpreting, seriating, inferring*

## For your information...

The body is a collection of simple machines. One type of simple machine found in the body is the lever. The type of lever is determined by the position of the fulcrum in relation to the effort and the load. When the arm raises an object, the elbow is the fulcrum, the hand holds the load, and the upper arm flexor muscle applies the effort. Similarly, the jaw, thumb, and forefinger are levers.

In this bag, students use levers as extensions of their bodies. They are asked to compare their thumb and forefinger to other levers. They observe the muscle movements and the control they have when picking up objects with their hands. Each movement is the result of nerves conducting electrical messages from the brain to the muscles, which tighten and relax. In contrast, manmade levers are much simpler in construction, more difficult to manipulate, and less precise. Students can come to appreciate the complexity and precision of their body. They can also appreciate the challenge faced by people who have to rely on a prosthesis in place of their arms and hands.

## Materials needed for the bag...

- levers (e.g., chopsticks, tweezers, ice tongs)
- marble
- penny
- balloon

## Words to know...

| | | | |
|---|---|---|---|
| lever | fulcrum | effort | load |
| prosthesis | tendons | muscle | simple machines |
| force | forefinger | | |

## Now try this...

- Research prostheses, such as the myoelectric arm or the Boston arm.
- Construct a device to pick up and move objects.

- Visit a rehabilitation center in a hospital.
- Research other simple machines.

**Look for these...**

*How Things Work* by Michael and Marcia Folsom. Collier Macmillan, 1987.

*How Sport Works* by The Ontario Science Centre. Kids Can Press, 1988.

*The Magic School Bus Inside the Human Body* by Joanna Cole. Scholastic, 1988.

*Teammates* by Peter Golenbock. Gulliver Books, 1989.

*Looking at the Body* by David Suzuki. Stoddart, 1988.

# Pinch and lift!

**You will need:**

- levers, such as ice tongs, tweezers, chopsticks
- several objects, such as a balloon, marble, penny, cup, book, chair

**What to do:**

Use the levers provided to pick up and move each of the objects.
Use your thumb and forefinger to pick up and move each of the objects.
Compare how the levers and the thumb and forefinger work.

**What to notice:**

- how the levers hold and move objects
- how the thumb and forefinger hold and move objects
- which objects are easier/harder to pick up each way
- what muscles are used each time
- how the muscles are used each time
- what other parts of the body are working at the same time

**More to do:**

Tape the thumb to the palm of the hand and try to write, turn the page of a book, tie a shoe, or button a coat.
Discover other parts of your body that are levers.
Research animals which have body parts that are levers.

# Overview — Easy moves it

THEME: *Motion*

SCIENCE AREA: *Physical*

SCIENCE SKILLS: *Experimenting, inferring, making a model, observing*

## For your information...

Friction is always present when one surface moves over another surface. The greater the surface area, the greater the friction, and the more wear and tear on the objects as the resisting force acts against the object's direction of movement to slow it down. Sliding friction, rolling friction, and air resistance are types of friction. Of the three, rolling friction creates the least friction because it involves the smallest surface area and reduces "rubbing" surfaces. The smoother the surfaces rubbing together, the less friction and consequently, the less wear and tear.

The application of rolling friction was known to Egyptians who used logs set under large blocks of stone to build pyramids. This technique reduced friction by limiting the contact points on the stone and ground's surfaces, making the movement of large objects easier. In this bag, the easy-mover device created is similar to the one the Egyptians used — the marbles are like the rolling logs and the objects are like the stones. The marbles used in the easy mover function in the same way as ball bearings. Modern technology uses ball bearings to change sliding to rolling friction, thereby significantly reducing heat and machinery wear.

As students pull objects over the surface without the easy mover, they experience sliding friction. When they repeat the task with the easy mover, they experience rolling friction. Their attention is drawn to the difference in force needed to move the objects.

## Materials needed for the bag...

- piece of cardboard
- marbles
- lid from a box (at least 20 cm by 20 cm)

## Words to know...

| | | | |
|---|---|---|---|
| rolling friction | sliding friction | bearings | Egyptians |
| pyramids | spring scale | technology | resistance |

**Now try this...**

- Find applications of roller bearings in everyday objects.
- Take apart different junk objects, such as a bike wheel or a skateboard, to find the bearings.
- Use a spring scale with and without the easy mover to measure both the static and kinetic friction.
- Research Newton's Laws.

**Look for these...**

*Inventors' Workshop* by Alan J. McCormack. David S. Lake Publishers, 1981.

*Physics for Every Kid: One Hundred and One Experiments in Motion, Heat, Light, Machines, and Sound* by Janice Vancleave. J. Wiley & Sons, 1991.

*Wheels* by John Williams. Wayland, 1990.

*1001 Wonders of Science* by Brian Williams and Brenda Williams. Random House, 1990.

*Power Magic* by Alison Alexander. Simon & Schuster, 1991.

*Black Inventors of America* by McKinley Burt Jr. National Book Company, 1969.

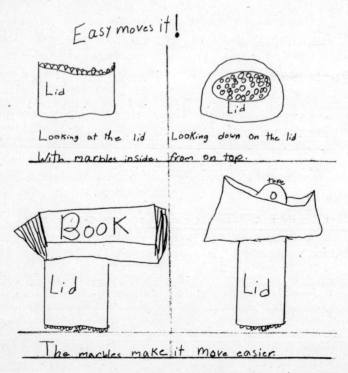

Easy moves it!

Lid — Looking at the lid with marbles inside.

Lid — Looking down on the lid from on top.

BOOK — Lid

tape — Lid

The marbles make it move easier.

# Easy moves it

**You will need:**
- medium-sized lid
- variety of heavy objects, such as a book, brick, or can of juice
- marbles
- piece of cardboard

**What to do:**

Fill the lid with the marbles.
Place the cardboard on top and flip the lid to create an easy mover.
Set the easy mover on a flat surface, such as the floor.
Make sure you can see the marbles under the rim of the lid.
Carefully slide the cardboard out from under the easy mover and move it around.
Take an object and move it across the floor.
Put the object on top of the easy mover and move it again.
Repeat the last two steps with different objects.

**What to notice:**
- how the easy mover moves
- how objects move quickly across the floor by hand
- how objects move across the floor on the easy mover
- which way is easier to move heavy objects

**More to do:**

Feel the objects after moving them across the floor.
Find uses for your easy mover around the house.
Invent another kind of easy mover using marbles.

# Overview — Go car go!

THEME: *Motion*

SCIENCE AREA: *Physical/technology*

SCIENCE SKILLS: *Experimenting, making a model, predicting, communicating*

## For your information...

Technology is a way of linking a scientific concept with a real-life experience. In this activity, students apply their understanding of how things work by building a model. They construct and experiment with a simple machine — a wheel and axle. They build a model using familiar materials — the car's chassis. They then power their vehicle with a form of energy — electricity.

This bag activity is complex and should be used with students in need of a challenge. Wheels and axles are difficult to make and stabilize. In order to ensure success in this activity, suitable materials should be prepared and provided by the teacher. The coat hanger wire, which is used for the axles, should be cut 4 cm longer than the length of the cans. Soup cans containing clear broth work best as wheels because they can be emptied while leaving the tops and bottoms intact. Poke drainage holes at either end of the cans. Since these holes are for the axles, make sure they are centered and aligned. Rinse and drain the cans. The vehicles can be dismantled when the bags are returned and the cans reused. If students have already made wheels, they can make their own wheels and axles.

A single-shaft motor can be purchased at an electronics or science store. The energy is transferred from the motor to one of the wheels via the elastic band. As the motor's shaft rotates, the elastic band around the can sets the car in motion. Attaching the elastic band to the front wheel creates a front-wheel drive because the wheel is pulling the car. Attaching the elastic band to the rear wheel pushes the car, creating a rear-wheel drive. Reversing the poles of the battery changes the direction that the shaft rotates. During their investigations of battery connections, students observe how electricity flows. This serves as the experiential base for them to design a switch similar to the one described here.

A switch can be made by putting an elastic band around the

battery. Make sure that the band touches both poles of the battery. Wrap an elastic band around the length of the battery and tape it in place around the battery width. This creates an on-off switch when one paper clip wire is left under the elastic at one end and the other wire is inserted and taken out as needed.

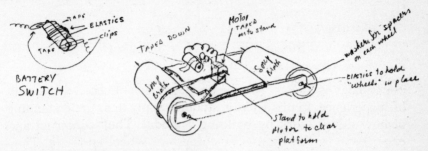

## Materials needed for the bag...

- 1.5 volt motor
- 2 paper clips
- piece of cardboard
- 1.5 volt battery

- 6 elastic bands
- scissors
- white glue
- 2 soup cans (broth)
- 4 washers

## Words to know...

rear-wheel drive    switch    poles    motor    shaft
front-wheel drive    axle    battery    electricity    chassis

## Now try this...

- Experiment with other materials for the wheels.
- Change the speed of the motor by changing gears.
- Research automotive transmissions.

## Look for these...

*Making Things Move* by Neil Ardley. Franklin Watts, 1984.
*Turning to Wheels* by Ed Catherall and Bev McKay. Chrysalis Publications, 1988.
*The Mice and the Clockwork Bus* by Rodney Peppé. Lothrop, Lee, and Shepard, 1986.
*Junk-Pile Jennifer* by John F. Green. Scholastic, 1991.
*Electricity and Magnetism* by Terry Jennings. A Templar Book, 1992.
*The Science Book of Electricity* by Neil Ardley. Doubleday, 1991.

# Go car go!

**You will need:**

- 1.5 volt motor
- 2 paperclips
- heavy cardboard
- 1.5 volt battery
- 6 elastic bands
- coat hanger wire
- scissors
- white glue
- 2 soup cans (broth)
- 4 washers

**What to do:**

Construct a car body or chassis with the cardboard.
Secure the motor and the battery to the chassis.
Connect one can to the motor's drive shaft with an elastic band.
Pass the axles through the cans and frame to create the front and back wheels.
Wrap an elastic band around the end of each axle to prevent slipping.
Connect the motor to the battery with the paperclips.
Hold the car in the air, place it on the floor, and let it go.

**What to notice:**

- how many wheels moved when the car was in the air
- which wheel drives the car
- what direction the car moves when placed on the floor
- how quickly the car moves

**More to do:**

Reverse the battery connections and observe the wheel movements.
Transfer the elastic band on the drive shaft to the other wheel.
Design a switch to turn the motor on and off.

# 5

# Connections

## Science/technology events

During the year, foster positive attitudes through events with a science/technology focus. Classrooms or divisions may hold introductory or culminating events as part of collaborative units. Current events may prompt information sessions open to parents and the community. Events could be in conjunction with or sponsored by parent-teacher associations. Suggested topics might include inventions, computers, building structures, environmental issues, and star-gazing. Other related topics include the participation of the physically challenged in science and equity issues.

Whether these events are held during the school day or in the evening, they are an excellent venue for a Sign Out Science component. For the most part, science/technology events feature speakers, demonstrations, storytelling, media presentations, and displays of students' work. Each of these is completed with an integrated hands-on activity.

That is where the bag activities come in. Participants of an "Inventors' Afternoon" could be presented with Sign Out Science bags containing an assortment of materials from which to invent useful devices. Those attending a "Star-Gazing Night" could be challenged by bags enabling them to identify and replicate constellations they see. Often there are pieces of software that enhance the focus of an event. When this is the case, the bags can be related to computer programs, such as *Earthquest Explorers Ecology* (1991), and used during an "Environmental Forum." In this instance, bags could also relate to the conditions affecting plant growth by setting up activities in which variables are identified, seeds are planted, and observations are made over time at home.

## Curriculum night

Parents like to feel they are part of their children's learning environment. Most schools set aside at least one time, usually early in the school year, to communicate with parents the philosophy and content of the curriculum. Many current educational initiatives and approaches differ from parents' own experiences in the classroom. Curriculum nights are opportunities for staff, administrators, and perhaps consultants to educate parents so they can be aware of the benefits their children derive from the program. Parents need to understand the reasons for changes and appreciate how these educational advances will impact on their children's future.

During curriculum nights, teachers outline the year's program, explain its relevance, and highlight special events. With an increasing emphasis on resource-based learning, teachers are looking beyond subject area boundaries. Their integrated curriculum reflects the overlap among subject areas. As discussed in Chapter Three, science can be infused into themes and found in other subjects. Parents are interested to learn how the science/technology content is being augmented in their children's classroom.

When Sign Out Science is being implemented, it represents a hands-on component to the curriculum. Teachers can explain why the bag program is used, how it works, and what they can do to help. Having class bags available for parents to explore puts them in touch with their children's experiences. Bags on a seasonal theme could relate to the focus of a particular curriculum night. Parents attending an evening in the middle term could experience a "Winter Wonderland" theme, and use the *Ice and snow* bags found in Chapter Four.

## Educational technology

Computer systems complement Sign Out Science. At the development stage, the content of some software can serve as a springboard for making new bags. Research done by computer includes accessing CD-ROM encyclopedias, commercial databases, and video- or laser-discs. A template can be formatted on computer. Students, parents, and staff can simply call up the template to design their own bags. A file may be set up in the form of a database to store research and instructions. Once saved on disk, bag collections can be manipulated efficiently. They can be sorted and classified by theme, science area, concept, or skill. Databases or the circulation software used in automated school libraries can be modified to assist borrowing routines at the implementation stage. During this time, students can maintain their science logs on computer using a word-processing package.

On-going communication about Sign Out Science activities is facilitated by aspects of computer technology. Word processing and desk-top publishing packages are useful for writing newsletters to send home. Whether part of a class, library, or school newsletter, or perhaps in a missive of its own, information about Sign Out Science could include tips, ideas, local science events, and places to visit. A new bag could be presented regularly in the form of an instruction sheet for which materials would be gathered at home. Lastly, with a modem, it is possible to exchange bag research and instructions with other schools near and far.

## Library/resource center

In our age of information literacy, the library-resource center, with its educational technology, becomes the hub of the school. This is, in fact, where the science-in-a-bag idea originated. Teacher-librarians are in an ideal position to use Sign Out Science in ways that are natural extensions of their role.

These educators are immersed in a world of material, human, and physical resources. Teacher-librarians orchestrate not only the practical, day-to-day functioning of the library, but also school-wide events and community contacts. It follows that they assume the role of coordinator for a Sign Out Science program operating from the library.

This would involve striking a committee of staff, parents, and experts that would develop the bags and implement the program. The library would be used to store materials, display resources, and circulate bags. The teacher-librarian would assist with parent volunteers, student monitors, and inter-library loans.

Teacher-librarians working with classroom teachers to plan collaborative themes draw upon a broad base of resources and their experiences from using these resources in a variety of ways. They have first-hand knowledge of the science/technology links in literature. When classroom teachers wish to integrate science into the curriculum by infusing Sign Out Science into their themes, the teacher-librarian can help redraft, design, and differentiate with bag activities. Although the program would be implemented in the classroom, teachers, students, and parents would find themselves in the library-resource center from time to time to research, access, and down-load bags by modem.

## Science buddies

Similar to reading buddies, science buddies pair older students with younger ones, within or across grades and divisions, to do Sign Out Science bags. Particularly effective pairing occurs when primary students are paired with intermediate students. Their science sessions can focus on bags from the library, the classroom, or their personal collection. Students can find bags they think their partners would like or ones that they favor and want to share. Given a purpose for choosing a bag, both partners benefit from the decision-making process, the chance to develop science concepts, and the opportunity to reinforce science skills.

In their role as facilitators, older students gain a sense of responsibility and leadership. They have the opportunity to be nurturing and patient with less advanced children. The time spent on science is increased for both, and the younger children have a supportive environment in which to develop their oral language skills. Younger students benefit from the extra attention, listening opportunities, and science experiences the buddy system offers. Their enthusiasm is well received by older students, who offer guidance and support their inquisitive nature and positive attitudes. Extending ranges of social interaction helps students, especially those without older or younger siblings, relate to different age groups.

Planning suggestions for a year-round science buddies program:
- select two classrooms, pair the children, and schedule sessions every two or three weeks
- use student-designed interest inventories to initiate the program
- have students record experiences and reflections in ongoing science response journals/logs
- present a range of bags related to class themes, individual interests, and science areas
- celebrate holiday and special events together for added social interaction
- produce individual or collaborative bags
- enjoy a special activity with refreshments to culminate the program.

## Science clubs

In any school, there are always a myriad of extra-curricular activities. Teachers with a special interest often take on these activities in the form of clubs. For example, science clubs are generally scheduled after school so that students can come together and pursue their interests. Members can invite guest speakers, such as a parent, other staff, an author, a scientist, or other students. Sometimes parent volunteers assist with the club. Once a regular schedule of meetings has been established, enthusiasm can be maintained by promoting the club in the school newsletter and announcing its accomplishments on the public-address system.

Science Club advisors may wish to:
- use Sign Out Science bags as activities for students during club time
- encourage club members to sign out the bags
- make up new bags with club members
- initiate and administer the program school-wide
- join other clubs occasionally, such as a computer club or a book club.

# Science fairs

Schools generally stage annual science fairs in a variety of formats. Traditionally, students research and present projects to be judged. A shift in thinking is taking place regarding the purpose and benefits of fairs. With more emphasis on projects related to the curriculum and less emphasis on competition, meaningful child-centered projects are gaining credibility. Students are encouraged to derive enjoyment from process-oriented learning. Personal experience provides the context for them to gain an understanding of science concepts. Simple projects, because of their familiar materials and limited variables, are relevant and within students' realm of understanding and control. Guest judges bring expertise and a fresh perspective to the event. Evaluation checklists for judges and students help everyone know the judging criteria, a factor that helps to focus attention on the process rather than the final product.

---

Sign Out Science fair tips for...
primary students:
- use existing bags as show and tell items for projects
- develop bag(s) as a theme-related class project
- encourage cooperative learning work groups to develop and explain projects.

early junior students:
- use existing bags plus a simple extension as show and tell items
- focus on understanding and explanations of extensions
- encourage independent or small work groups to develop and explain projects.

late junior students:
- use existing bags to show in-depth extensions or create new bags
- show research steps and describe how scientific method was used
- demonstrate interrelationships between bags and science in everyday life.

---

Sign Out Science bags are process-oriented learning projects that can have a place in science fairs. Once students have signed bags out, they can begin to understand the process and develop bags of their own. Sign Out Science bags provide students with a format that follows the scientific method which they can use in preparing their science fair projects. Knowing the format allows them freedom to design experiments and test science concepts. Recognition could be in the form of adding bags from the science fair to the school's or class's collection.

## Science Olympics

Teachers can take their cue from the world of sports to offer students a change of pace in the way science is approached. A morning, afternoon, or evening can be set aside to mount a science Olympic event. This cooperative group learning promotes science and could even tie in with an athletic theme. The organization of a science Olympics day is similar to a sports day of play events. Select a variety of bags that have short challenges and involve timing of events, such as *Speedy marbles*, page 65. Students are divided into teams of four or five. They may enjoy creating banners for their team and cheering to boost morale. The bags are placed at various stations around the room. The route through these stations is explained and posted.

The Olympics begin with one team positioned at each station. Students complete the bag activity and then rotate among the remaining stations. A time limit may be set forth for each station to keep the flow moving smoothly. The bag activities should engender a light-hearted, creative, and humorous tone; remember that the purpose is to have fun with science. The recognition each team and its individual participants receive upon completion should reflect this goal. "Prizes" may be awarded, not only to the teams who have finished first, but also to those who have finished last, were the loudest, the quietest, the messiest, and the funniest. Perhaps "dignitaries" (parents whose work is related to science, science teachers, etc.) could be on hand to present the medals.

Variations on Sign Out Science Olympics are limitless. They may have themes, such as solutions or motion. When they are held outside, the events take on a more recreational flavor with seasonal themes, such as bubbles, water, and camouflage in the warm months. Students who have arranged visits with their buddies, pen pals, or twin schools could host a science Olympics for these friends. There would be added excitement in participating in groups of peers differing in backgrounds and locations.

## Pen pals and twinned schools

Children benefit from relating to peers outside their immediate school environment. Maintaining pen pals within the community, nationally, or internationally has long been popular. Often educators who meet at conferences and other occasions discover they teach the same grade and are interested in having an authentic motivation for their students to communicate with others. Writing to pen pals fits into all areas of the curriculum because students recount stories about life at home and at school. Whether the class has a new member, is celebrating carnival, or has visited a museum, pen pals want to know and are interested in reading about it. Often hand-made cards and small craft gifts are exchanged. Sign Out Science bags are also items to share with pen pals. Students may write about their favorites, down-load stored bags, research, and instructions, or mail the bags and wait for a response. In this way, bag activities become the focus for shared communication between students who are experiencing science in a social context.

It is also possible to "twin" classes to establish communication links. Some countries have formal programs that match classes through a central organization. Teachers may find "twins" through informal networks. The emphasis tends to be on class-to-class contact rather than individual pen pals. By exchanging information packages during the year, classes learn about the culture, social studies, and environment in another part of the world. Copies of the Sign Out Science bags used in a classroom could be included in the information packages to send to their twinned class. How exciting it would be to receive bags made by the twinned class in return. It is fascinating to compare bags developed by students from their respective cultures. In these ways interaction with others helps students acknowledge their differences and celebrate their similarities.

# Professional resources

Allen, D. *Hands-On Science: 12 Easy-to-Use High Interest Activities for Grades 4-8*. West Nyack, NY: Simon & Schuster, 1991.

Bachor, D.G. and Crealock, C. *Instructional Strategies for Students with Special Needs*. Scarborough, ON: Prentice-Hall Inc., 1986.

Bosak, S.V. *Science is...* second edition. Richmond Hill, ON: Scholastic, 1991.

Burnie, D. *Reader's Digest How Nature Works*. New York, NY: Reader's Digest, 1991.

Cameron, C., Politano, C., and Morris, D. *Buddies: Collaborative Learning through Shared Experience*. Madison, WI: Creative Curriculum Inc., 1989.

Carletti, S., Girard, S., and Willing, K. *The Library/Classroom Connection*. Markham, ON: Pembroke Publishers Ltd./Portsmouth, NH: Heinemann Educational Books Inc., 1991.

Gamber, R., Kwak, W., Hutchings, M., and Altheim, J. *Learning and Loving It*. Portsmouth, NH: Heinemann Educational Books Inc., 1988.

Hann, J. *Reader's Digest How Science Works*. New York, NY: Reader's Digest, 1991.

Heimlich, J.E. and Pittelman, S.D. *Semantic Mapping: Classroom Applications*. Newark, DE: International Reading Association, 1986.

Lyman, F.T. "Think trix: A classroom tool for thinking in response to reading." *Reading Issues and Practices*, 4, 1987.

Knight, J. Evans. "Coding journal entries." *Journal of Reading,* 34, 1, September, 1990.

Osborne, R. and Freyberg, P. *Learning in Science: The Implications of Children's Science.* Portsmouth, NH: Heinemann Educational Books Inc., 1985.

Parsons, L. *Response Journals.* Markham, ON: Pembroke Publishers Ltd./Portsmouth, NH: Heinemann Educational Books Inc., 1990.

Petersun, R. (Ed.) *Innovations in Science.* Toronto, ON: HBJ, Holt, Rinehart and Winston of Canada, 1991.

Sprung, B., Froschl, M., and Campbell, P. B. *What Will Happen If...: Young Children and the Scientific Method.* New York, NY: Educational Equity Concepts Inc., 1985.

Williams, B. and Williams, B. *The Random House Book of 1001 Wonders of Science.* New York, NY: Random House, 1990.

Willing, K.R. and Girard, S. *Learning Together: Computer-Integrated Classrooms.* Markham, ON: Pembroke Publishers Ltd., 1990.

Winzer, M. *Children with Exceptionalities,* second edition. Scarborough, ON: Prentice-Hall Inc., 1990.

# Index